THE MENTAL AFTERMATH

The Mental Aftermath

The Mentality of German Physicists
1945–1949

KLAUS HENTSCHEL

University of Stuttgart, Germany

OXFORD

UNIVERSITY PRESS

OXFORD

UNIVERSITY PRESS

Great Clarendon Street, Oxford OX2 6DP

Oxford University Press is a department of the University of Oxford.
It furthers the University's objective of excellence in research, scholarship,
and education by publishing worldwide in

Oxford New York

Auckland Cape Town Dar es Salaam Hong Kong Karachi
Kuala Lumpur Madrid Melbourne Mexico City Nairobi
New Delhi Shanghai Taipei Toronto

With offices in

Argentina Austria Brazil Chile Czech Republic France Greece
Guatemala Hungary Italy Japan Poland Portugal Singapore
South Korea Switzerland Thailand Turkey Ukraine Vietnam

Oxford is a registered trade mark of Oxford University Press
in the UK and in certain other countries

Published in the United States
by Oxford University Press Inc., New York

© Klaus Hentschel 2007

Translation of *Zur Mentalität deutscher Physiker*
in der frühen Nachkriegszeit 1945–1949 by Klaus Hentschel
originally published in German by Synchron Publishers, Heidelberg 2005

Translated by Ann M. Hentschel

The moral rights of the author have been asserted
Database right Oxford University Press (maker)

First published in English 2007

All rights reserved. No part of this publication may be reproduced,
stored in a retrieval system, or transmitted, in any form or by any means,
without the prior permission in writing of Oxford University Press,
or as expressly permitted by law, or under terms agreed with the appropriate
reprographics rights organization. Enquiries concerning reproduction
outside the scope of the above should be sent to the Rights Department,
Oxford University Press, at the address above

You must not circulate this book in any other binding or cover
and you must impose the same condition on any acquirer

British Library Cataloguing in Publication Data

Data available

Library of Congress Cataloging in Publication Data

Data available

Typeset by Klaus Hentschel in Garamond Font using LATEX 2e
Printed in Great Britain
on acid-free paper by
Biddles Ltd, King's Lynn, Norfolk

ISBN 978–0–19–920566–0

1 3 5 7 9 10 8 6 4 2

CONTENTS

LIST OF FIGURES

"Historical mentality is the ensemble of sages, thought, and sensibilities distinguishing a specific collective at a specific time." (Peter Dinzelbacher, 1993)

1

INTRODUCTION

Mentality, according to Jacques Le Goff, is that which a person shares with other individuals, that which, as it were, subliminally allows or inhibits thoughts, words, deeds, and feelings.[1] Not being something a historical actor is normally conscious of, we cannot rely on contemporary self-analyses or reflections alone in order to gauge a given mental climate. Contemporary sources must therefore be examined with a different eye, in order to reconstruct it and see beyond the original intent. We must read between the lines—something we are already accustomed to doing with texts from the Nazi era. Mentality expresses itself not only in factual statements but also in apparently innocuous, spontaneous side remarks or formulations, repetitive turns of phrase and themes.[2]

Conventions and academic customs are not what concerns me here, nor contemporary ideologies or worldviews.[3] Neither is what Bourdieu would call the "habitus" of German physicists of the immediate postwar period the focus of this study: that very consciously adopted figure of authority, the masculine hero and obedient servant of the discipline and the state.[4] I seek to probe the intellectual climate of the time at a deeper, more inaccessible level. We might call it the collective psychology which comprehends the outlook, emotional state, and disposition toward certain actions. I specifically chose the word "mentality"

[1] See Le Goff [1974] p. 81: "La mentalité a vaincu la doctrine. Ainsi ce qui semble dénué de racines, né de l'improvisation et du réflexe, gestes machinaux, paroles irréfléchies, vient de loin et témoigne du long retentissement des systèmes de pensée." See also ibid., p. 86, regarding "discours obligé et machinal."

[2] Gerd Tellenbach's definition (1974) is along the same lines: "Mentality is natural, self-evident, indeed even impulsive behavior and response, casual thinking and opinion, scarcely operating at the conscious level." (Cited in Sellin [1985] p. 559, with further references on conceptual history of mentality.) Instead of "behavior," I would prefer "comportment," to specify a tendency or disposition towards certain behavior.

[3] Theodor Geiger (1932) draws the line between mentality and ideology: "Mentality is, metaphorically speaking, an atmosphere—ideology is a stratosphere. Mentality is one's skin—ideology one's robe." (Cited in Raulff (ed.) [1989] p. 10.) The focus here is on inextricable mental dispositions of entire groups of people.

[4] Examples of this attitude among scientists in the KWG/MPG are given in Schüring [2006] pp. 9, 258f., 281; likewise among mandarine scholars of the *Kaiserzeit* in Ringer [1969]. On the habitus concept in general, see Bourdieu [1993].

as part of the title of this study because the focus is less on specific actions than on collectively perceived horizons and expectations, on hopes and fears. Historians of the French *Annales* school have been developing the history of mentality since the 1930s. Marc Bloch refers to "représentations collectives" or "l'atmosphère mentale," Lucien Febvre to "l'outillage mental," and Ariès to "sensibilité collective" or "attitudes." J. Revel and Gourevitch specify "catégories générales de représentation," Le Goff distinguishes "le quotidien et l'automatique," and Vovelle, Certeau, and Faye refer to "l'imaginaire." In allusion to Cassirer's "Weltbilder," Mandrou and Duby describe a "vision du monde." Such collective dispositions channel notions, attitudes, and ideas in two ways: *positively* by guiding individuals within the group in a particular direction, without there being any outright coordination or explicit agreement among them; or *negatively* by blocking out or repressing other conceivable options that would appear astonishingly self-evident to uninitiated outsiders. These restrictive and resilient components of a mentality are underscored by the historian Ernest Labrousse as "résistance" and by Robert Mandrou as "kinds of reticence." Fernand Braudel goes so far as to refer to a "prison de longue durée." German historians have added further layers. Hagen Schulze, for instance, defines mentality as "the field of notional climates, of social norms and axioms, of collective perceptual worlds and patterns of legitimization, gauges of 'right' and 'wrong' comportment—in brief, a collective subjective reality as a decisive causal prerequisite for social actions just as, in the normal case, for individual actions."[5]

My methodological approach is remarkably close to the one used by the historian Frank-Michael Kuhlemann in Bielefeld in his social study on the mentality of Protestant vicars in Baden between 1860 and 1914.[6] The present book likewise examines a prosopographically sharply defined group with its own specific socialization and environment, and hence a "micro- or particular mentality" that finds its place within the larger framework of "macro- or total mentalities." The micro-mentality distinguishes itself by a susceptibility to specific influences granting it a

---

[5]For surveys on the history and methodology of the history of mentality see, e.g., Duby [1961], Le Goff [1974], Vovelle [1979], Burke [1986], Schulze [1985], Sellin [1985]. Le Goff's article "Histoire des Sciences et Histoire des mentalités" in the special issue of *Revue de Synthèse* [1983], defends an imprecise and malleable concept of mentality as advantageous, because it can be heuristically modified to suit a given issue. Tellenbach [1974] pp. 11ff. regards this as a major drawback of this method. On Bourdieu's concept of "habitus" as a "system of structured and structurizable dispositions" or (somewhat more usefully) as a "subjective but non-individualized system of internalized structures, common schemes of views, ideas and approaches," see Bourdieu [1993] pp. 97ff., 112.

[6]For the following see Kuhlemann [1996] pp. 186f., [2001] pp. 37–46, 189–203.

relatively narrow "latitude of thought and action." Here, too, a "group-specific notional fabric" is worked out that establishes the "interpretational horizons" of the actors, which, as dispositions, form a "transmission belt between a given environment and its resulting specific living standards." Just as, in Kuhlemann's case, *Bürgerlichkeit* and religion become "mentality themes" around which the discourse of his group coalesces, so also, in the present study, are Russian phobia, reconstruction, and responsibility of scientists, to mention only a few.

The history of mentality frequently focuses on extended structures *de longue durée* that sometimes persist for centuries. Le Goff even defines the history of mentality as a history of slowness.[7] My approach is rather oriented toward Georges Duby's, who distinguishes temporal, spatial, and social layers of mentality ("différentes cadences") and also considers shorter cyclical transformations in mentality.[8] As in Michel Vovelle's or Frank-Michael Kuhlemann's analyses, with the French Revolution or World War I, respectively, triggering such an "upheaval of mentalities," two historically contemporary events also demarcate the period under investigation here: the end of World War II in 1945 and the founding of the two German republics in 1949.

Mentality does not change overnight, of course. But during this brief period it evolved relatively quickly as a result of altered defining conditions in German politics. For this reason these five postwar years are an extremely interesting transition phase containing *relics* of a waning mentality along with *germs* of its successor. National Socialism had not yet completely vanished from the public discourse as it had, say, in Adenauer's Germany. Nor had it degenerated into the "antifascist" ritual later to establish itself on the other side of the iron curtain. But the externally imposed denazification procedures fed an emotional bias into the equation. Unconscious linguistic usage and certain norms and values lived on in the postwar period, such as regarding the role of women or the expected character profile of an aspiring physicist. The minutes of the Swiss Federal Polytechnic's school board in Zurich offer a concrete example. Since its founding in 1855, the polytechnic often had a very high proportion of Germans among its faculty. In a confidential discussion by the board in 1950, one federal councillor thought it would not be politically wise "if we nominate a German at almost every reappointment. Germans, it is known, cannot easily change their mentality."[9] Thus it is not simply

---

[7] See Le Goff [1974] p. 82; cf. Dinzelbacher [1993] p. XXVI and Schulze [1985] pp. 256ff.

[8] Georges Duby [1961] pp. 948ff. On deep changes of mentality: Tellenbach [1974] pp. 22ff., Kuhlemann [1996] p. 200.

[9] The reaction by School Board member Gotthard Egli to the federal councillor of the Swiss Social Democrats, Willy Spühler, was: "If [the low-temperature technician Peter] Grassmann

a construct of malevolent historians to note a persistence of a chauvinistic and authoritarian mentality beyond the collapse of the Third Reich. Contemporary reports already reveal it: Hartmut P. Kallmann (1896–1975) had been dismissed from his position as department head at Haber's Institute of Physical Chemistry in 1933 and deprived of his permission to teach at university because three of his grandparents were of Jewish origin. Kallmann decided not to leave Germany and worked under the protection of IG Farben's president on a private fellowship. Although he was rehired in 1945, becoming director of the Kaiser Wilhelm Institute of Physical Chemistry, he ultimately decided to leave Germany in 1948. In December 1947, Michael Polányi (1891–1976) reported to Robert Havemann (1910–1982), a physicist in East Berlin: "Kallmann is determined to leave Germany, where he feels terribly unhappy both on account of past persecutions, and of the still existing traces of nazi mentality in the German milieu."[10] The specialist in applied mechanics from Göttingen Kurt Hohenemser (1906–2001), and many other emigrés or otherwise ostracized persons, complained about undiminished Nazi mentality: "the way things are going in western Germany [...] in far too many places the prevailing spirit is not so much national socialistic properly speaking as nationalistic and anti-Semitic."[11] The primary motivation behind Kallmann's eventual departure for the US in 1949 was his disgust with vestiges of uneradicated Nazism.[12] The present study will focus on typical attributes of the specific mentality among physicists during the first postwar years. Longstanding cultural and socio-psychological traits of an all-encompassing German mentality cannot be addressed here in detail. A few prominent features in the discourse of the

should distinguish himself by the known German tone, I, too, would rather advocate a rejection, but not merely because of his past [as a member of the SA]." Quotes from *Schulratsprotokolle der Eidgenössisch-Technischen Hochschule*, meeting of 24 Jun. 1950, p. 216, cited in Gugerli et al. [2005] pp. 242f., 437.

[10]M. Polányi's letter dated 9 Dec. 1947 from the files of the *Society for the Protection of Science and Learning* (Bodleian Library, Dept. of Western Mss., no. 479), where Polányi plied for aid for Kallmann, is cited in Deichmann [2001] p. 484. There is unfortunately no file copy of Polányi's original application among his papers.

[11]K. Hohenemser to R. Courant, 21 May 1946 (I am grateful to Gerhard Rammer for referring me to this letter in the collection at the Manuscript Division, Library of Congress.) For similar complaints by Hermann Broch about persistent Nazi mentality (more closely defined below and, e.g., in Benz [1990]), see Broch [1986] pp. 14f., 61, 87, 121, 134. Durkheim points to the resilience of bygone mentalities in general: "For is not contained in each of us various doses of the person from yesterday?," cited in Bourdieu [1993] p. 105. See also Vovelle [1979] pp. 19, 30ff.

[12]Kallmann accepted a position at the US Army's Signal Corps Laboratories in Fort Monmouth, New Jersey. He was appointed professor of physics at New York University in 1950. On Kallmann's remarkable character see Spruch [1965], Schüring [2006] pp. 214f.

postwar period included, for instance: arrogance and aggressiveness in a position of dominance, submissiveness in a weak position; an exaggerated masculine ideal restricting the feminine realm to cradle, kitchen, and church; an obsession for titles with its attendant minority complex; a yearning for a powerful state and strong-arm leadership; a tendency toward introspection; and a compulsion for scapegoating.[13] If we reach a little further back in time, we find strains of what the historian Fritz Ringer has masterfully portrayed as the "mandarin" mentality among German academics during the Kaiserreich. The Jewish theoretical physicist Ludwik Silberstein (1872–1948), who had obtained his doctorate in Berlin, moving subsequently to London in 1913 and to Rochester, New York, in 1920, draws an apt picture of this "Prof.-Geheimrath-Deutschland" in his attempt to lure Albert Einstein away from the ever bolder anti-Semites in Germany:[14]

> Mrs. Einstein did tell me (in Princeton), you had a kind of moral "duty" (a perfectly mystical concept in the present case) not to abandon the Germans just now, "who have lost virtually everything, you know." Nonetheless, I am deeply convinced that Germany (by this I mean the atmosphere of German profs., privy councillors, court councillors, and the like—as the working class in Germany is also free from Junkerdom and other such nonsense) is not the right place for you. The Lenards, Gehrckes, etc.—their names are legion—are perhaps all (with the sole exception of Planck and the deceased Rudolph Virchow) petty and at the same time brutal individuals, Junkers and at the same time pitiful slaves of the Kaiser's regime.

Long after the abdication of the Kaiser and the defeat of Nazi Germany, this mandarin mentality remained very much alive among German *Ordinarien*, as the obituaries of many academic physicists and institute directors in postwar Germany reveal.[15] Older mentalities are not simply displaced by younger ones but overlain, their contours occasionally remaining discernible through them for a while.

The overall scheme of my study is thus a plastic layering of mentalities, deeply embedded layers overlain by intermediaries, which themselves are topped by others only valid for individual groups and for shorter periods of time.[16] This topmost

---

[13]For interesting observations along these lines, see esp. Balfour & Mair [1956] pp. 53f. The grim image of German culture and mentality depicted by Rose [1998] is unacceptably lopsided, as pointed out in Hentschel [2000].

[14]See L. Silberstein to A. Einstein, 4 Sep. 1921, Collected Papers of Albert Einstein, archive no. 21 046. On Silberstein, see Duerbeck & Flin [2005]; on the German mandarins, see Ringer [1969].

[15]For examples see, e.g., the selection of short portrayals of key figures of the former polytechnic in Stuttgart after 1945, in Becker & Quarthal (eds.) [2004]. There are similar more detailed biographical studies on Sommerfeld, Pohl, Gerlach, Bothe, etc.

[16]Vovelle regards mentalities "comme stratifiées, objet d'une histoire [...] à étages": Vovelle [1979] p. 34. Benz [1990], Arendt [1993], and Balfour & Mair [1956] delve into these deeper layers.

layer is our central theme. That is why this study is explicitly confined to the group of German physicists who did not emigrate. I point out here that the concept of mentality (as documented, e.g., by the quote by Kallmann) is an actor's category. That is, the historical figures used it themselves to encapsulate the complex set of dispositions, moods, and attitudes guiding their actions.

I do not intend to pass self-righteous moral judgment from the safe vantage point of retrospection. Nor is this intended as an apology for commissions and omissions of the period. The aim here is a neutral description of the prevailing mentality.[17] If we can better understand *how* people of that time thought and felt, we can better understand *why* they acted and wrote the way they did. Let there be no doubt about my own position. It was one of the most depressing experiences I ever had as a historian to see reflected in the documents how very soon *after* 1945 the chance of coming to grips with the National Socialist regime was allowed to slip by, dismissing the opportunity to make a frank assessment of the facilitating conditions the regime had set.[18] Initially incomprehensible behavior is often the subject of analyses in the history of mentality. This applies as much to history as it does to ethnology—which some leading historians of mentality liked to compare themselves to.[19] It was for precisely this reason that historians studied the mentality of the Middle Ages and the early modern period. Our topic is an interesting test case for whether effective short-term analogues can be found to Braudel's "prisons de longue durée" in a modern setting, from which an explanation for the compulsive refusal to confront the Nazi past might emerge.[20]

[17] On the problem of neutrality see, e.g., Balfour & Mair [1956] p. 5: "if he succeeds in preventing himself from expecting pigs to fly, [he] is apt to sound as if he pitied the animals for their inability to do so."

[18] I share this dismay with Walter Jens [1977] pp. 342f. After going through the files of the University of Tübingen, he declared: "the minutes from the era after 1945 were the spookiest: It was as if nothing had happened! No Stalingrad and no Auschwitz, no eugenic sterilizations and no academic ennoblement of anti-Semitism! [...] No process of grieving, no admission of guilt, no taking stock, no self-reflection in the spirit of an examination of one's conscience." This text is cited in Ute Deichmann's excellent study on chemists, [2001] esp. p. 451.

[19] This probably explains the predominance of studies on medieval mentality as opposed to early modern and recent episodes. See, e.g., Le Goff [1974] pp. 77f. The pejorative connotation of the term mentality, which is irrelevant to us here, is partly explained by its usage in anthropology. Lévy-Bruhl used it to describe supposedly primitive "sociétés inférieures": ibid. pp. 84f. Weart [1988] also avoided the word mentality in the title of his history of attitudes toward nuclear energy. This alienation was felt even by contemporaries, as we see from Karl Wilhelm Böttcher's article in *Frankfurter Hefte* 4 (1949) p. 492: "Germany has become in many ways an 'unknown land,' even for us Germans. Our observations and experiences everywhere confirm this fact."

[20] Other episodes in the history of physics and science that have already been studied from the

Parallels to Jacques LeGoff's observations about "historical mentality" can also be drawn with the attitude of physicists in the aftermath of the Nazi regime. Here the test persons to be examined for "a society's relationship with its past in its collective psychology" are not historians, with whom LeGoff preferentially exemplified it, but German physicists.[21] We shall see that this group is an excellent indicator of the "sentiments a country's population has about its past." Physicists have a tradition of relative frankness and open communicativeness. After 1945 they also held key positions as science policy-makers and planners in East and West. This special role among scientists in general was augmented after the atomic bombs were dropped over Hiroshima and Nagasaki in August 1945. Public opinion and the scientific community attributed this feat to physicists—at the expense of their many collaborators in engineering and chemistry.[22]

The focus here is not on biographical descriptions of individual perceptions. Our view will reach beyond the personal level to aspects valid for many individuals at once. It will circumscribe a "space or horizon typical of the period, the 'expressive frame,' within which comportment is formed, while allowing a certain leeway for individual deviations."[23] One individual perhaps lingered long on the summit of self-justification, while another remained sunk in the valley of self-pity. Constant vacillation between the extremes, within the complex mental landscape to be charted here, was more often the case, however. It is not possible to go into the details about the many orientational points lending this landscape its topology: the ordinances and laws imposed by the Allies, the wretched working conditions in many places after the war, etc. But it is obvious that a mutual relationship existed between the prevailing mentality and the immediate living conditions.[24]

point of view of the history of mentality include the chauvinist declarations by scientists during World War I (the "war of the intellectuals") or the attitude of physicists with respect to acausality during the Weimar Republik (Forman's thesis).

[21] See LeGoff [1992], esp. pp. 167ff. on historians as "mediators of this collective opinion" about the past, who (just like our physicists) had an influence on the opinions of other groups.

[22] The special role physicists played in the postwar period is discussed, for instance, by Carson [1999], Badash [2003], and Beyler [2003] (also in the discourse of others, e.g., Karl Jaspers and Martin Heidegger). Schirrmacher [2005b] follows this trend into the early Federal Republic of Germany.

[23] Dinzelbacher [1993] p. XXII. These must be distinguished from a historical "radius of action" as defined by Vierhaus [1983]. On the tension between individual and social psychology, see also "Histoire et psychologie" (1938) in Febvre [1953/65] pp. 207–220.

[24] Vovelle [1985] is a model study of this mutual relationship between mentality and social or political processes within the relatively limited time frame of the French Revolution.

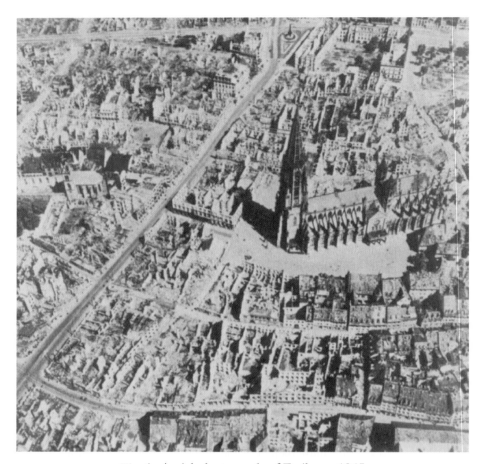

Fig. 1: Aerial photograph of Freiburg, 1945

The carpet bombing of German cities by the Allies had destroyed one fifth of all living quarters and two fifths of all transportation facilities. The centers of major cities had been reduced to rubble. 1.85 million apartments were completely uninhabitable, 3.6 million damaged. At the end of 1946 only 8 million living quarters existed for 14 million households.[25] As an illustration, 31 cubic meters of rubble were consigned to each surviving inhabitant of Cologne. In Berlin the rubble was estimated to total 55 million cubic meters. Theo von Laue (1916–2000) got a taste of the local black humor when his father Max von Laue (1879–1960) wrote him: "What's Berlin called nowadays? 'Greater Caputh-on-the-Spree' [a

---

[25]The above statistics are taken from the excellent online exhibition by the DHM: www.dhm.de/lemo/html/Nachkriegsjahre/DasEndeAlsAnfang/kriegszerstoerung.html For other documentary images, see e.g., Schaper et al. (eds.) [2002] pp. 6ff.

pun on a lakeside vacation spot west of Berlin ('kaputt' means broken)]. What's Lichterfelde-West called? 'Trichterfelde—Rest' [i.e., 'crater field—remains']."[26]

Fig. 2: Street life in Berlin after the capitulation, 1945. Source: DHM (LEMO) http://www.dhm.de/lemo/forum/kollektives_gedaechtnis/011/ index.html

Squalid living quarters in overcrowded, make-shift housing, a dire scarcity of food and clothing, freezing cold in the wintertime, and a constant influx of more dispossessed refugees: these typical elements have to enhance the mental picture I draw.[27] Ernst Brüche made note in his diary on 3 May 1945 of the comparatively generous weekly rations of 250 g meat, 1.5 kg bread, 1 l low-fat milk, and 2.5 kg potatoes, remarking "I am quite ready to eat up these atrocious amounts in a single day. Stretched over a whole week, it's barely enough to avoid starving." At the beginning of the following year the rations were cut down even

---

[26]Max to Theo von Laue, 16 July 1846, AMPG, div. III, rep. 50, suppl. 7/7, sheet 28. Beyler in Walker (ed.) [2003] p. 98 discusses the amounts of rubble and its symbolic dimension.

[27]Living conditions during that time are described, e.g., by Balfour & Mair [1956] pp. 113ff. Heidelberg, Göttingen, and a few other places were spared from the bombing. About the working conditions of physicists at universities and research institutes elsewhere, see *PB* **2** (1946) no. 1, pp. 23f., no. 2, pp. 14f., no. 3, pp. 65–70, no. 4, pp. 85–89, no. 5, pp. 116–121, etc. Merritt & Merritt (eds.) [1970] pp. 15–19, 191f. discusses the main worries of the population.

further:[28] In December 1946 one of the coldest spells in central Europe started and food reserves were rapidly depleted, prompting major hunger demonstrations and strikes in the industrial Ruhr region and other major cities in March 1947. Only in July did the tense situation gradually relax again, but things only really began to get better again in the summer of 1948.[29]

Fig. 3: Entrance of the bombed-out Institute of Physics at the University of Frankfurt, 1945. Source: Müser [1989] p. 157.

The Frankfurt Physics Institute was in a sorry state, as we learn from the following report. The building had lost its windows and shed its plaster, its piping leaked, and its roof had partially collapsed, but it still stood on firm foundations:

> The director's room at the Physics Institute had no doors, no window panes and some of the walls had even lost all their plaster, but the walls themselves were sound, so the room could be converted into a make-shift home. But it was too large for the purpose. So a kind of wooden hut was built inside the room with a surface area of about 6 square meters and two

---

[28] Brüche's diary no. III (BLM, box 7) and no. V, 19 Jan. 1946. About Hannover and Hamburg see diary no. VI, 12 Nov. 1946. M. von Laue wrote L. Meitner on 27 Nov. 1947 about the destruction, clothing shortage, and consequences of malnutrition; Lemmerich (ed.) [1998] p. 309.

[29] See http://www.dhm.de/lemo/html/Nachkriegsjahre/DasEndeAlsAnfang/hunger.html and Marshall [1980] pp. 659ff. See, e.g., supplements by F. H. Rein in AMPG, div. III, 14A, no. 5730 dated 1946. *PB* **4** (1948) pp. 42f. published a declaration by the Technical University of Munich on the nutritional situation signed, among others, by Ludwig Föppl, F. von Angerer, and G. Hettner. For intellectual workers it demanded "extra ration coupons at least at the level of semi-hard laborers."

meters in height. So in 1945, right after the war ended, this was the institute director's office, living room, and sleeping quarters, furnished with a camp bed, a small table, and a chair. As soon as gas supplies were reinstated, this cabinet could even be heated: with a Bunsen burner; when it was extremely cold, frugal Czerny even allowed himself a second.[30]

The polytechnic in Stuttgart fared no better. About 70 percent of its buildings were destroyed, much of its equipment and installations were no longer usable, and the remainder was confiscated by the Allies. Books and teaching materials had either been burned or removed to distant storage facilities. The Physics Institute had lost its windows and heating, and the floors of the laboratories were buried under a layer of rubble 20 cm high. The students were required to help clear it away as a precondition for admission. In all, it cost about 250,000 hours of student labor to restore the main building and some of the institutes, including the lecture halls and laboratories for physics and physical chemistry and the Institute for Materials Testing. Only the Institute for Technical Physics in the Degerlocher woods, where it had been relocated, and part of the Second Institute of Physics escaped the bombings.

When in August 1945 the cosmic-ray specialist Erich Regener (1881–1955) decided to return to the chair he had been ousted from by the Nazis in 1937, teaching in physics could resume in Stuttgart under his directorship. The conditions were far from satisfactory, of course. There were not enough rooms and insufficient equipment. The students had to put up with severe overcrowding (the average working space per student had shrunk from about six square meters in 1930 to half a square meter). Empty stomachs (the per-person intake of calories was 850 in 1947 against 2,800 in 1930) and a lack of books and working utensils added to the ordeal. A student's residence was, on average, five kilometers away (against two previously), 15 percent of Stuttgart's students were refugees, and about 45 percent financed their own studies with part-time jobs. Their average age, which in 1930 peaked sharply at 21–22, broadened out in 1947 to between 21 and 26 years, with an upward tendency (to 37 years). It took some time for these figures to fall off again.[31]

---

[30]Müser [1989] p. 157. The experimental physicist Marianus Czerny (1896–1985) directed the institute from 1938 until his retirement in 1961.

[31]On the above see Ringel & Grammel (eds.) [1947], esp. pp. 8ff. and app. for images and statistical data. Regener is discussed in Becker & Quarthal (eds.) [2004] pp. 18f., 302ff. See also E. Regener's letter to A. Einstein, 13 Sep. 1945; copy from the Collected Papers of Albert Einstein (call no. 20 016) in UAS, SN 26. For some postwar photographs of the ruined polytechnic in Stuttgart see here Figs. 6 and 7, p. 21.

Fig. 4: A crowded chemistry lecture at Stuttgart Polytechnic in 1947. Notice the advanced age of most of the students (photograph by Dessecker). Source: UAS.

Fig. 5: A seminar on general relativity in 1947 (photograph by Dessecker). Source: UAS.

The proportion of refugees was far greater elsewhere, such as in Göttingen, but otherwise these statistics are quite typical of the time. "Food and fuel shortage, occupation hazards, etc.," that was Max von Laue's description of the situation in 1946 in a nutshell.[32] James Franck's secretary painted a desperate picture reflecting its impact on the psyche:[33]

> How I and most people in Germany are getting along nowadays? The daily struggle for the bare necessities of life consumes every last bit of energy and time. Although living conditions are becoming more primitive by the day, an inordinate effort is necessary for the simplest things that one never used to have to attend to. There is no room left for intellectual or artistic pursuits. Life is exhausted by the constant grinding worry about sheer material existence. Added to that, there are unfortunately many points of friction with one's surroundings, because people are incredibly crowded together here and everybody is nervous and irritated.

In this study we will encounter representatives of virtually all the subdisciplines of physics, both theoreticians and experimental physicists, industrial and academic scientists. Their ages range from veterans of the First World War to beginning students. My focus is on the mentality of physicists who remained in Germany. But the perspective of their discriminated colleagues, who had been expelled and in many cases remained unwelcome even after 1945, is useful too. Their distance as emigrés is a true mirror—albeit not a plane one—of what was said and thought in Germany after 1945. Toward the end of my discussion I address attempts at communication between the two groups, which more frequently than not failed.

If there is indeed such a thing as *the* mentality of German physicists of the time, it should manifest itself not just among the great but also among the small. A few unknown figures, both young and old, will have their say besides more famous physicists. It is worth bearing in mind the following motto that appeared in the *Physikalische Blätter* in the laudatio of Arnold Sommerfeld for his 80th birthday:[34]

> Intelligence has a quantum-like distribution. Every day we encounter it, average quanta, small—and in the great majority—very small intellects converging on nill. If one may trust an important new book, when the Earth and humanity are gone they will still be there suspended in a cloud of stupidity in space. They reflect off us, we hardly feel the repercussion, the interaction is practically zero.

---

[32]Max to Theo von Laue, AMPG, div. III, rep. 50, suppl. 7/7, sheets 2–4, undated, c. Feb. 1946.

[33]Frau Schmidt to James Franck, August 1947; Franck papers, RLUC, box 8, folder 4; von Laue [1949b] p. 138 provides a survey of the staffing and general situation of physics in Germany in 1949.

[34]E. Buchwald: 'Arnold Sommerfeld zum 80. Geburtstag,' *PB* **4** (1948) pp. 456–459, quote on p. 457. See also Vovelle [1985] pp. 93ff. on the category "the masses" and their collective mentality.

I shall leave open to what extent the mentality of German physicists differs
from that of other scientists, members of other disciplines in the humanities and
social sciences, indeed the population as a whole. This could ultimately only
be answered by comparison against similar studies on the mentality of other
scientific communities, which unfortunately do not seem to exist yet (at least for
the postwar period).[35] A coexistence of various mentalities at the same time
would be conceivable, but I would think a subset model more probable, with
there being only minor deviations from the mentality of chemists or engineers,
for instance.[36] A cursory glance at Ute Deichmann's major study on chemists
under National Socialism reveals many points of agreement with the mentality
profile of physicists examined here. Her last chapter covers the period after
1945. Be this as it may, the documentation assembled here stems almost entirely
from physicists, categorized here by membership, short-lived though it may have
been, in the *Deutsche Physikalische Gesellschaft*. Hence those rejected by the society
in 1938 who explicitly decided not to rejoin after 1945 are also among them.
Testimonies by members of other disciplines are included where they express
something of general validity also reflected in such professional journals as the
Physikalische Blätter. A convergence of characteristics in the mentality of physicists
from numerous diverse sources will decide whether it makes sense to refer to
a specific mentality of German physicists. I contend that the following points
indicate such fields of convergence:

- Tension with the Allies
    - (a) superficial admiration and, occasionally, opportunistic familiarity
    - (b) covert reserve and distrust
    - (c) stubborn resistance to Allied control, e.g., denazification
- Russian phobia
- A sense of isolation and grief over the fragmentation of Germany
- Bitterness about the "export of scientists"
- Scapegoating the *Deutsche Physik* movement

---

[35] Ringer [1969] describes the mentality of academic "mandarins" in the period 1890–1930, based
primarily on examples taken from the humanities. Harwood [1993] and Harwood & Johnson et al.
in vom Bruch & Kaderas (eds.) [2002] juxtapose the Weber ideal of the specialist or expert, more
typical of scientists or technologists of the postwar period. Arendt [1993] provides a description of
the mentality of the German population in general that is as caustic as it is applicable.

[36] On the coexistence of various mentalities see, e.g., Le Goff [1974] p. 88. Duby [1961] proposes a
model of a more continuous mentality with slight group-specific and class-specific variations. Burke
[1986] reflects on the resulting tensions, assigning an aporetic place for the history of mentality
"between an intellectual history with society left out and a social history with thought left out."

- Forgetting, in dual form: on the one hand as amnesia and unconscious repression, on the other as reticence and dissimilation (conscious withholding or falsifying of information)
- Shame, listlessness, and lethargy remedied by workaholism
- Self-justification and the guilt issue
- Self-pity, sentimentality, and selfishness
- "Propaganda-free day-to-day" and political apathy
- New awareness of a scientist's responsibility
- Side-lining of emigrés and "soilers of the nest" (no "asymmetric discretion")
- Insensitivity in communicating with emigrés

Michael Balfour and John Mair identified three of these points as early as 1956: self-pity, sentimentality, and selfishness are pinpointed as general characteristics of vanquished nations.[37] Their report on the Allied occupation of Germany for the *Royal Institute of International Affairs* is a balanced description of the attitudes on both sides, also outlining the Germans' expectations at the time. Another characteristic given there (lack of objectivity) is omitted here because that probably generally applies to actors in any given historical situation. Ute Deichmann called attention to the characteristic of ubiquitous forgetting and repression as well as to an insensitivity toward emigrés in the last part of her book on chemists in the Third Reich. Hannah Arendt noted during her visit to Germany in 1950 the self-pity, listlessness, and inability to express emotions, the escapism, moral perplexity, and the refusal to grieve.[38] Her observations were based on a cross-section of the general population. The same traits emerge among my sampling from the physics world, indicating that it did not differ much from other groups. A few other facets covered in the various sections below have also been touched upon in various articles focusing on individuals, institutions, or episodes in physics, chemistry, engineering, etc., but by no means restricted to the sciences, in postwar

---

[37] See Balfour & Mair [1956] p. 63 in the section "Attitude of the Germans": "the sentimentalism, the selfpity, the selfishness, and the lack of objectivity which tend to characterize a nation in defeat." Balfour was a trained historian and was a member of the British branch of the Allied Control Council between 1945 and 1947.

[38] See Arendt [1993] pp. 24–28, 46. The report was published in 1950 by the American Jewish Committee but the German translation, by Eike Geisel, had to wait four decades to appear, in 1993. For a more thorough analysis of Arendt's thoughts about the postwar period and for biographical background information on her see Pilling [1996]. Others reached similar diagnoses. See Lifton [1986] p. 442: "psychic numbness"; Santner [1992] p. 153: "massive trauma"; Brumlik [2001]: a refusal to feel compassion; and Sime [2006] p. 37: "self-deception."

Germany. I would like to mention in particular the biography of James Franck currently being prepared by Jost Lemmerich, as well as Klaus Schlüpmann's study on Hans Kopfermann, a draft of which was available to me prior to its publication. Einstein and Heisenberg are not representative cases; neither are Wilhelm Müller and Johannes Stark. So the studies by David Cassidy and Cathryn Carson, Freddy Litten, and Andreas Kleinert were not very helpful in my search for a *typical* mentality profile of the time. But I do return repeatedly to Lutz Niethammer's and James F. Tent's denazification studies as well as to Anikó Szabó's impressive prosopographical review of the expulsion, return, and "restitution" (*Wiedergutmachung*) of university teachers in Lower Saxony. Some of the published studies by the Presidential Commission on the Historical Research of the Kaiser Wilhelm Society under National Socialism were also a valuable resource for postwar issues. This applies particularly to Michael Schüring's recently completed study on the Max Planck Society's postwar treatment of formerly expelled members. In conclusion, the "indispensability of a historical study on the mentality of West German postwar society" is pointed out.[39]

The present study offers a pioneering integrative synopsis of these points to stake out the mental landscape in which all the actors unconsciously moved, each negotiating their own way. To understand the slight variants of this mentality among individuals, perhaps even among subdisciplines, we must also consider the specific local differences, with their varying degrees of freedom defined according to the age, status, and self-consciousness of each actor, along with personal predispositions to particular patterns of thought and behavior. Even if no single postwar German physicist can meet this ideal on all counts here, the whole enterprise is therefore not necessarily discredited. From discussions in general history we are already familiar with the tension existing between a history of mentality oriented toward collective phenomena and the isolated biographical understanding. Ernst Hinrichs reflects: "Mentality as a category for the collective, the unconscious, and the long-term poses difficulties for a science that even today takes essential elements of its identity from the specifically German tradition of historicism, in which individuality and development play a prominently significant

---

[39] See Schüring [2006] p. 47. In concrete terms, its treatment is summed up as a "refusal to feel compassion" (*Einfühlungsverweigerung*), and verbal strategies of "semantic reconstruction" (*semantischer Umbau*) and apologetics.

role."[40] This applies at least as much to the history of science, which has had even less success at distancing itself from biography. Let us give it a try here anyway.

About the sources used

A main source of this study is the journal *Physikalische Blätter* (PB).[41] The hundreds of complete articles, shorter notices, and necrologies in its first few volumes constitute a serial source as employed in the French tradition of writing the history of mentality.[42] Its broad scope beyond that of a professional journal makes it particularly appropriate for our purposes. Originally conceived as a "physico-political journal," it was alluded to jestingly by its editor, the specialist on electron microscopy Ernst Brüche (1900–1985), as the "paraphysikalische Blätter."[43] This "informal noticeboard" accepted brief announcements by all the universities and colleges, published letters to the editor covering topics of the day, and also reprinted articles from other publications (some in excerpt or translated by Brüche). According to Brüche's foreword in its first issue, the *Neue Physikalische Blätter* (as it was first called in 1946) was also intended as a forum for chemists and mathematicians, teachers and engineers. One of them, Arnold Sommerfeld, was incidentally not impressed by it. He wrote to the publisher Vieweg in early 1946: "May the 'Physikalische Blätter' expire, they are not quite worthy of your publishing house."[44] Notwithstanding this scathing verdict, the *Physikalische Blätter* attracted a large readership, becoming "a physicist's book of epistles [*Hauspostille*]," thanks to its up-to-date and multifacetted reporting. Thus it is a useful source for the mentality of physicists. Even Sommerfeld changed his mind about it. In 1949 he wrote to the editor: "Your Blätter are always so interesting that I would like to

---

[40]Hinrichs [1979/80] p. 226. "Trying to draw the physics of an entire nation under a single umbrella seems to be [...] an unpromising enterprise," writes Gerhard Rammer in his review of a book (unfortunately based mainly on secondary sources and hence not useful for my purposes) by Gabriele Metzler: *Internationale Wissenschaft und nationale Kultur. Deutsche Physiker in der internationalen Community 1900–1968*, Göttingen, 2000. See the journal *NTM,* new ser. **9** (2001) p. 132.

[41]The relevant German *PB* article titles appear here in translation, with more specific referencing in the annotation. Square brackets signal the years of publication of more generally relevant entries listed in the bibliography.

[42]For examples of this research approach, see Thiriet [1979/80]; for some criticism of it, see Reichardt [1979/80] pp. 235f. I subscribe to the assertion: "tout est source pour l'historien des mentalités" by Le Goff [1974] p. 85.

[43]This formulation occurs, e.g., in a notice by Brüche in BLM, box 6, folder 66.

[44]The history of the *Physikalische Blätter* is outlined by Dreisigacker & Rechenberg [1994], [1995] pp. F139ff. Hermann [1995] pp. F102f. discusses Brüche's original intention. For a history of the *Deutsche Physikalische Gesellschaft* after 1945, see Walcher [1995]. Regarding Sommerfeld's letter to Vieweg from 24 Jan. 1946, see DMM, collection 89, 014.

subscribe to it, despite my abhorrence of reading piles."[45] We must bear in mind, however, that these articles as well as those from the *Göttinger Universitätszeitung* (*GUZ*) or the *Neue Zeitung*, for instance, could only appear after receiving the stamp of approval of the military government. So the image it reflects is of the official side of the coin.

That is one reason why we have to supplement it with excerpts from contemporary correspondence. Exchanges of letters between close colleagues are a finer gauge of the mood of a period. They are a window into the true thoughts and emotions among colleagues. The following were particularly useful exchanges: the papers of Bonhoeffer, Hahn, von Laue at the Archive of the Max Planck Society (AMPG), the Born collection in the Staatsbibliothek zu Berlin Preußischer Kulturbesitz (SBPK), Gerlach's papers in the Deutsches Museum in Munich (DMM) or Brüche's papers at the Landesmuseum für Technik und Arbeit in Mannheim (BLM) as well as the papers of James Franck, Eugene Rabinowitch, and Michael Polányi in the Regenstein Library at the University of Chicago (RLUC) and the Wergeland papers at the Trondheim University library. The archive of the Deutsche Physikalische Gesellschaft (DPGA) currently set up in the Magnus-Haus in Berlin proved useful regarding administrative issues. Among my many various pickings I would like to point out Michael Eckert's extremely useful annotations of the Sommerfeld correspondence at the Deutsches Museum, which is also accessible via the Internet: www.lrz-muenchen.de/~Sommerfeld, and Jost Lemmerich's publication of the Meitner/von Laue correspondence. Other selected editions focusing on specific historical issues, such as the Pauli correspondence, were also very useful for this study. (For the sake of brevity, German quotes will only appear here in translation. The originals are available in the German edition of this study.)[46]

A mosaic of the mentality of German physicists is thus pieced together from scattered sources in the hope that a somewhat representative picture will emerge out of a comparison of the contemporary documentation from various pens and various points of view. There is no guarantee, of course, but there certainly are a few touchstones. The findings tend to point in a common direction. A few existing synthetic sources already draw such comparative considerations. Examples are the summary reports by a few Allied Control officers as well as

---

[45] A. Sommerfeld to E. Brüche, BLM, box 2, folder 20 and likewise 9 May 1947: "I read your 'Blätter' with the greatest enjoyment and shall continue to do so" (box 119, folder 308). In a congratulatory speech in honor of Brüche's 70th birthday, Sommerfeld even complimented his journal as the "Blätter, which means the world" (box 6, folder 66).

[46] See Hentschel [2005].

by a delegation of the British *Association of Scientific Workers* and the *Institution of Professional Civil Servants*. They traveled throughout Germany in 1948 in order to determine whether "German scientists are aware of the extent of their complicity in the ruinous machinations of the Third Reich and what they are doing to give their activities a new ethical orientation."[47] Another excellently informed group of observers of Germany and the Germans after 1945 had been assigned by the *British Intelligence Objective Subcommittee* (BIOS) to travel through the occupation area to ascertain the level research in university and industrial laboratories had reached during the war. Many of these officers were themselves self-made specialists; in some cases they were emigrés. The results of their interviews were eventually documented in almost 2,000 reports that were then released to companies at home only too eager to exploit the free know-how of their direct German competitors. These *BIOS Reports* also give a good idea of the mentality and basic attitudes of German scientists. We start with a selection of such synthetic impressions.[48]

---

[47]See Hahn [1949] p. 2, Murray [1949b]. Beyler & Low in Walker (ed.) [2003] p. 111 discuss the British Association. The *Federation of Atomic Scientists* was founded in the USA on motivations similar to the Democratic Scientists' Association (abbrev. *Minka*) in Japan.

[48]On the following see Reid [1976] pp. 259–265. It casts other reports in the shade. Kuiper [1946], for instance, outlines his visit with German astronomers between April and September 1945, hardly mentioning the mentality of his colleagues. His report is almost exclusively confined to current research projects. Likewise the report by University Education Control Officer Geoffrey Bird [1978], in charge of the Univ. of Göttingen. General impressions of the German attitude are provided by the *OMGUS Surveys* 1945–49, compiled by Merritt & Merritt (eds.) [1970]; see also Marshall [1980] on the British zone, Arendt [1993], and Scherpe (ed.) [1982]. The international eye-witness reports in Enzensberger (ed.) [1990], esp. pp. 79ff., provide a European comparison to Germany.

2

SCIENTISTS IN GERMANY SEEN FROM THE OUTSIDE

Quite a few of the Allied observers asked themselves the question still open today about whether scientists in Germany should be regarded as collaborators or as victims of the Nazi regime: "Should one equate them with the conscious plotters against society, or with the dumb, if frequently complacent, 'followers'?" According to one unidentified BIOS officer in an article about the fate of German science during the Nazi regime, published in the British journal *Discovery* in August 1947, their involvement in the military structure of the German Reich was too great simply to pass them off as little cogs of the system. The privileged position physicists enjoyed, their leaning toward authoritarianism, and their willingness to cooperate on the development of weaponry like the V-2 rocket, as well as in the plundering of laboratories in the occupied territories, all speak for themselves. The role of German scientists before 1945 does not concern us here, however. The general mood after 1945 was described as "extremely susceptible to despair and defeatism which, if allowed to decline into isolationism, will prove as good friends to reaction in the future as in the past." Scientists were portrayed as continuing to be of the opinion: "Science and politics do not mix," even though the devastating and sometimes criminal consequences of this ideology were everywhere strikingly visible.[49]

> The German scientist seems to be drifting in a social vacuum, without con-
> tact with other classes. His personal position is dangerous and precarious,
> with the reduced scope for scientific work under the level-of-industry plan,
> even if raised 50% above the present: large numbers of the most highly
> qualified men are looking outside Germany for employment. [...] their first
> hope is the United States; as terms of employment are apparently better
> over there than in Britain, British employment comes a poor second in their
> calculation, though it is still more desirable than employment in Germany.
> None apparently, contemplates making a career in other European coun-
> tries, though short-term employment is sometimes undertaken. There is
> basically a keen desire to get out of the Europe they have helped to ruin.

[49]On the foregoing see Anonymous [1947], quotes on pp. 239f., 243. The periodical *Discovery* was affiliated with the *Association of Scientific Workers*. For similar statements, see Murray [1949a & b] as well as here p. 120. A virtually complete set of the BIOS *final* or *miscellaneous reports* is available in the Crerar Library of the University of Chicago. The Staats- und Universitätsbibliothek Hamburg unfortunately has an incomplete set.

Fig. 6: The gutted main building of the Stuttgart Polytechnic in 1947 was soon reconstructed and still marks the center of the downtown university campus. Source: UAS (photograph by Dessecker).

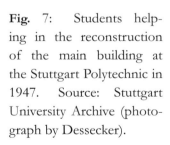

Fig. 7: Students helping in the reconstruction of the main building at the Stuttgart Polytechnic in 1947. Source: Stuttgart University Archive (photograph by Dessecker).

The purpose of Richard Courant's (1888–1972) trip in 1947 for the *Office for Naval Research* was primarily to evaluate German progress on the development of calculating machines at a number of universities and technical colleges. At the same time, he was on the lookout for promising young researchers. Courant had been a mathematician at Göttingen until 1933, later moving to New York. His reconnoitering of the general situation among university teachers and their students serves us well as a kind of mood barometer. After landing at Frankfurt he and his company were amazed at how difficult it was to find one's way around in the labyrinthine ruins of the city, and "at the demoralization of the population. They were shaken by the sight of great crowds of ragged, hungry Germans, many of them begging."[50]

At the polytechnic in Darmstadt, Courant observed that the registered students, numbering over 2,000, all put in a half day every fortnight toward clearing and reconstruction work in the totally bombed-out campus. This condition of their matriculation was common elsewhere as well. Its president Vieweg had complaints about the young helpers, not only because of their underfed condition but also because of their poor academic quality and lack of any sense of ethics: "for some, it seemed that the only thing the Nazis had done wrong was to lose the war."[51] At Heidelberg, Courant perceived "a slight lack of resonance" in a conversation with the publisher Ferdinand Springer. We shall encounter this peculiar incommensurability between the emotional and linguistic worlds of resident versus exiled German. At Marburg the mathematician Kurt Reidemeister (1893–1971) told Courant that the age group between 20 and 25 were too poorly educated to come under consideration for his purposes. Those between 30 and 35, however, had another drawback: political *Belastung*: "thus, carrying the stigma of having been National Socialist Party members—but he is sure that many of the belastet have not mentally been Nazis. [They] had to join the Party in order to keep the positions... [Still] he said that 90 per cent of the population, including academic people, are dangerously but not hopelessly nationalistic. In natural science people in general, much less, however."[52]

---

[50]Reid [1976] p. 259. Cf., e.g., Balfour & Mair [1956] pp. 11ff. on the extent of the destruction in Germany.

[51]Reid [1976] p. 260. See also the assessment of Göttingen students by Bird [1978] p. 149: "Many were still dazed by disillusionment, a few were arrogant and unfriendly, but all were keen to get down to their studies."

[52]Reid [1976], pp. 260f. (orig. English). A delegation of the *Association of Scientific Workers* that visited various university and industrial laboratories in 1948 came away with a more extreme impression after seeing German scientists and speaking with Allied Control officers. "The Control

The impression his fellow Germans at Göttingen left in July 1947 was not favorable: "absolutely bitter, negative, accusing, discouraged, aggressive."[53] An interesting dichotomy between Werner Heisenberg's (1901–1976) superficial and inner mental profiles was revealed in conversation. During his first few interviews with Courant he was prepossessing and basically positive, as he had been before the war. But that was evidently a façade reserved for outsiders. He was completely different once he had reassured himself that it was safe to speak frankly:[54]

> But another day, discussing politics, he found that Heisenberg came out finally with the same stories and aggressiveness against Allied policy of starvation [evidently meaning the dismantling of German factories] as the less cool and more emotional people.

As this example shows, communication failed both at the level of superficial friendliness between colleagues as well as at the emotional level. Their basic moods and values were far too far apart. Courant wrote a friend: "I found very few people in Germany with whom an immediate natural contact was possible. They all hide something before themselves and even more so from others."[55] All the same, Courant's final verdict at the end of his visit in 1947 was that the Germans definitely had to be helped. He wrote to Warren Weaver (1894–1978) of the *Rockefeller Foundation*:

officers were of the opinion that all Germans they had come into contact with were the same nationalists as before. As one of them put it: 'If they could only invent a weapon that would destroy the whole occupying force without lifting a hair of any Germans, they would be jubilant.'" (Cited by Murray [1949a] p. 171.)

[53] Reid [1976] p. 261 (the following passage: ibid., p. 265). See also Rudolf Ladenburg to Lise Meitner, 15 Dec. 1947 (Churchill College Archives, MTNR 5/11, sheet 45R, quoted by Vogt [2002] p. 128): "there seem to be only very few left with whom one could communicate well, at least among physicists. Courant & Frau Artin were alone in Göttingen for 2 weeks & spoke with pretty much everyone there and in Hamburg, Darmstadt, Marburg, etc., whom we used to know well from earlier." This letter is quoted further here on p. 156.

[54] Reid [1976] pp. 261f. For Heisenberg's basic political stance see, e.g., Cassidy [1992] pp. 331ff., 501ff. and Heisenberg's first postwar letter to Arnold Sommerfeld, Sommerfeld papers, DMM 1977-28/136-43.

[55] Courant to Winthrop Bell, cited by Reid [1976], p. 263. The conversation with the mathematician Helmut Hasse (1898–1979) in Berlin was similar: "Met Hasse. Mixed feelings." (ibid.). Mitscherlich & Mitscherlich [1967] p. 35 gives a psychological interpretation of these "communication problems between Germans and the rest of the world after the war had ended." See also the introduction to the published correspondence between the Austrian writer and emigré Hermann Broch (1886–1951) and the diplomat and publicist Volkmar von Zühlsdorff (born 1912) in Broch [1986] pp. 12–15 and their letters on pp. 23ff., 128f.

> In spite of many objections and misgivings, we feel strongly that saving
> science in Germany from complete disintegration is a necessity first because
> of human obligations to the minority of unimpeachable German scientists
> who have kept faith with scientific and moral values.... It is equally necessary
> because the world cannot afford the scientific potential in German territory
> to be wasted.

The aerodynamicist Theodore von Kármán came to a quite different conclusion. After an equally thorough series of official visits of research sites in ruined Germany, he had a much more pessimistic view of the ability of Germans to change their ways for the better. At the end of April 1947 he expressed his opposition to any kind of further involvement by the *Rockefeller Foundation* in Germany, in no uncertain terms. Weaver noted in his diary:[56]

> He [Kármán] thinks that at least 80 per cent of all the present German
> faculties and German students are completely unrepentant and arrogant.
> He says that if we do one thing for them, we will simply justify their own
> opinion of us as fools. When I ask him what he thinks we ought to do with
> them, he shrugs his shoulders and says: "Just leave them alone for about
> fifty years."

Sources such as these are very instructive. From an aggregation of numerous individual observations and conversations, some witnesses of the time managed to arrive at a higher plane and give a summary description of the general mood among their contacts.

---

[56] Entry of 25 May 1947, Rockefeller Archive Center, RG 12.1, diary of Warren Weaver, p. 45, cited in Schüring [2006] p. 325. On Kármán's life and work see his autobiography Kármán [1968]. There was a flood of—often contradictory—recipes for what to do with Germany after the war: see Stone [1947b].

3

TENSIONS WITH THE ALLIES

(a) Superficial admiration and opportunistic friendliness

Michael Balfour, a historian working for the British branch of the Control Council, interpreted the surprising "submissiveness in defeat" of the Germans, so contradictory to the image of the enemy, as the extreme psychological opposite to their arrogance and aggressiveness in a position of strength.[57]

A turncoat of this type from among our physicists is Wilhelm Westphal (1882–1978). He tried to charm the new men in power into employing him despite his clearly incriminating political record. Lise Meitner (1878–1968) wrote about this to James Franck (1882–1964) in early 1946:[58]

> Right after the arrival of the Russians in Berlin, he wrote an article in a newspaper in which he portrayed himself as Planck's protector and a true Democrat. After the Hitler regime had begun he became a Nazi, not just a superficial one, which one would have ascribed to a lack of courage, but a "convinced" Nazist, hanging a big portrait of Hitler in his home. And you know that after the first World War he was such an avid Democrat as to denounce a German Nationalist colleague and was discharged from the Ministry as a consequence. If a person adjusts his convictions three times according to the prevailing political wind and then tells me that he has a good conscience, he has shown his true colors and needs no further comment. He simply has no conscience. It is very unfortunate that he managed to persuade the naive Americans by his methods, but I hope he won't persuade you as well.

---

[57] See Balfour & Mair [1956] p. 53. See there pp. 58f. about alternative attempts to play the Allies off against each other by feeding bits of information to one occupying force about misconduct by the other. The emigré monetary scientist from Kiel Gerhard Colm described the "coolie-like spinelessness of his German counterparts" after a preparatory visit in the spring of 1946 pertaining to the currency reform: Krohn in vom Bruch & Kaderas (eds.) [2002] p. 444. Hartshorne noted the "disgustingly complacent attitude taken by German personnel," in Tent (ed.) [1998] p. 37.

[58] L. Meitner to J. Franck, 16 Jan. 1946, Franck papers, RLUC, box 5, folder 5 (corrected to 1947 by Sime [2006] p. 33, note 159, as Meitner's misdating). See also Franck's undated draft letter to Meitner (ibid.), about the "distress" Westphal's assurances about his "clear conscience" were causing him. "I fear he will have to bear some disappointment. But, then again, if he didn't, I would be dismayed about the lack of judgment on the part of the local military and other official authorities."

This was not even an extreme case. In Meitner's view Westphal was quite typical. She continued: "The majority of the Nazi leaders were people of Wilhelm West-phal's stamp, with no inner guiding thread, no conscience, and without the least whisper of truthfulness. These people are and were the bane of Germany." (ibid.) Even scientists known for their integrity became inconsistent out of exaspera-tion. When Otto Hahn's (1879–1968) negotiations with Colonel Bertie Blount (1907–1999) about the renaming of the Kaiser Wilhelm Society ran aground, he threatened to change sides, warning that "scientists, who had always considered the help of the English or Americans positively, might turn their sights on France or Russia. I doubt whether this would be in the interest either of the English or the Americans."[59]

Numerous notices in the *Physikalische Blätter* acknowledge the competency of their new rulers outright, as far as technological accomplishments and scientific know-how are concerned. Examples include advances in color television in the USA (in its second volume of 1946, p. 21, based on a report in *Die Neue Zeitung*, 7 Jan. 1946) or the detection of hydrodynamic rolling waves of the earth's surface during tests of the atomic bomb (on p. 49, taken from an article in *Time*, 28 Jan. 1946), or the radio-controlled igniter design deployed against the V-1, attaining a downing rate of 70 percent by the end of the war (on p. 20, from *Yank*, 25 Nov. 1945). The awe-inspiring total of 3,800 collaborators on American radar development is quoted, roughly 700 of them being physicists: "twice as much as had worked on the atomic bomb" (ibid., p. 18), etc. This last example illustrates the ambiguity of many of these compliments. An undercurrent of envious defensiveness is often perceptible. Sommerfeld's praise of the "wonderful organization" with which the US had secured its advantage in nuclear energy implies a justification for the failure of the German effort. What a convenient explanation for Germany's inadequacy! In a letter to an officer of the occupying force, this doyen of theoretical physics in Germany tried to make political capital out of this flattery:[60]

> It is hard for me to advise you on the control of nuclear energy, because the question is so closely connected with the current political situation. I fully understand that the USA do not want to lose their lead in the exploitation of nuclear energy, which they gained by their wonderful organization, but I believe that no other country can catch up for years to come.

---

[59] A note in the files about a conversation between Blount, Hahn, and Telschow on 10 July 1946, MPG Archive, founding files 4, cited in Walker (ed.) [2003] p. 41.

[60] A. Sommerfeld to Oliver Johnson, 5 Sep. 1947; www.lrz-muenchen.de/~Sommer feld/Kurz Fass/04452.html

Regarding the question of atomic education, I would like to refer to the hybris in Greek tragedy. Wise management is set against it, which achieves more in the long run than boundless egoism. Humanity will have to learn this anew, if it is not to perish from the hybris of atomic energy.

A reviewer of the book *Science at War* praised the apparatus in Rotterdam similarly in 1949 as the high point of British radar development:[61]

The authors point out emphatically that this astonishing development was primarily the result of uninhibited and intimate collaboration among all the participants. The problems were pursued, ignoring rank and distinction, in discussion and critique between Air Force marshals, admirals, scientists, pilots, laboratory assistants, development and production engineers.

To every German reading between the lines it was clear that the author was decrying the lack of such "uninhibited and intimate collaboration" between German researchers and the military. The almost dictatorial power of institute patriarchs like Pohl in Göttingen and gray eminences like Sommerfeld in Munich hardly changed after the war until as late as the 1960s. The structure of physics departments at American universities, on the other hand, led to a much stronger democratization, at least within the faculties.[62]

Brüche's notice in the *Physikalische Blätter* entitled "Flying eyes" indicates how strongly the war continued to govern the visions of the future:[63]

The battlefields of World War III will know no secrets. Flying televisors will watch how cities disintegrate and they will project the fighting onto the generals' viewing screens in deep bunkers. Even bombs will observe their own fall to the target with unperturbable electronic eyes, before report and bomb vanish simultaneously.

Many texts by German physicists from 1945 and 1946 try to present the military defeat as a rationalization for the failures of German science and technology.

---

[61]Jetter: "Science at war," *PB* **5** (1949), no. 1, p. 45. A later example is the comparative juxtaposition of the editions for R & D in the USA and Germany in Brücher & Münster [1949] pp. 335f.

[62]Maier-Leibnitz once said to me, quoting Theo Mayer-Kuckuck: "Departments are an extension of the *Ordinarius* principle to the level of assistants." Horst Korsching's grumblings in the Farm Hall transcripts about Gerlach's dictatorial authority is symptomatic of the German academic structure.

[63]"Fliegende Augen," *PB* **2** (1946), p. 71, taken from *Time*, 1 Apr. 1946. Atypically, this opportunity to emphasize German achievements in developing televisor controlling devices prior to 1945 was not made use of. Arendt [1993] p. 29 mentions an "undertone of satisfaction" in conversations among Germans about the next war.

By pushing the responsibility entirely on politics they found quick absolution. An apparently unpublished manuscript among the papers of the editor of the *Physikalische Blätter*, entitled "Our new task" is typical of this genre:[64]

> What we feared has come to pass. All the efforts and privations our nation undertook were in vain. They were necessarily in vain, because one of the decisive factors was recognized too late: technology as a factor of power and science as the foundation of technology. There can be no doubt that the scientific and technological superiority of the opposing side decided the outcome of the war.

After a litany about the loss of the submarine war and command of the skies over Germany, it continues:

> we had to realize that the opponent had succeeded in getting and using shorter wavelengths than ours and that they had more efficient, new position-finding equipment. That was not the only thing that determined our defeat but it may have been enough, even if other mistakes and unlucky circumstances hadn't compounded it. German physics and technology was not up to date and thus the question poses itself: Are we scientists at fault? This question must be answered with a resounding No. The cause of the failure is ultimately a too low regard for science in Germany, which unmistakably persisted in recent decades.

What strikes modern-day readers is the nonchalant way in which the "efforts and privations of our nation" are linked to the strategic goals of the National Socialist leadership. The author's diaries and correspondence even prior to 1945 portray him as quite distant from the Nazi ideology. But he seems oblivious of the potential repercussions of a success in military physics and technology in Germany. It would have propped up the criminal regime and might well have extended the war, perhaps even changing its course in Hitler's favor. Instead of being relieved that this is not what had happened, Brüche laments the missed "chances" and the resulting subjugation to the Allies, who are perceived as "opponents," not liberators.

---

[64]"Unsere neue Aufgabe," undated two-page typescript with corrections and additions, BLM, box 5, folder 58; see also there a two-page typescript by Brüche "On assessing my opinion and attitude toward National Socialism" (Zur Beurteilung meiner Einstellung und Haltung gegenüber dem Nationalsozialismus), a multi-page text: "I and the party. Experiences of a German physicist from 1933 to 1945" (Ich und die Partei ...), a number of denazification forms, diaries, and various other supporting documents for Brüche's hearings before the trial court in Mosbach in December 1947, BLM, box 123, folder 321.

Ernst Brüche's private reflections in his diary about the differences between American and German soldiers are highly interesting:[65]

> All in all, I have got an image of Americans as a rich nation of high technological standards. Proud, unapproachable, and in everything technically superior and efficient. [...] We are sluggish, chase after ideals that in reality are completely different from what we think and we don't even notice. We have a tick for exactitude and don't let ourselves be convinced that the others have long since found a simpler way that might not stand up to German criticism but leads more quickly to the goal and has been followed with success.

And on 28 June 1945, as negotiations with an American major were faltering:

> We don't understand the Americans and they don't understand us. [...] We will have to continue to strive and work, so that they see that all of us had not been Nazis and that it is for their own good that they don't commit the same error with the Germans that we committed with the Jews.

This misdirected comparison between National Socialist persecution of the Jews and Allied treatment of the Germans after 1945 is repeated a number of times in Brüche's diary. It exposes how lopsided his assessment of the situation was. When the continuing negotiations about such essentials as petrol rations and a spare tire for his motorbike were proving difficult, Brüche grumbled in his diary about the Americans:

> These people remind me somehow of playing children, giant children, who thanks to their great strength have occasion to play with the Germans. It is cat playing with mouse. Does the cat realize at all that it is hurting the mouse when it allows the mouse, half dead as it is, to run a little more for its dear life so that it can catch it again? Why this disinterestedness by a nation that has taken upon itself the responsibility along with the power? Is this a game or cold calculation by the leadership? We want to work and rebuild. Why don't they allow it? Why aren't the trains running yet? Why is the post unusable and the telephone line broken? The Americans have been here for four months, four months of "peace"—and we are still waiting for peace. We are living off capital. Raw materials and supplies are everywhere lacking.

---

[65] Ernst Brüche: diary no. III, 22 Apr. 1945 and no. IV, 28 and 30 June and 27 Aug. 1945 (BLM, box 7). On 7 Nov. 1945 Brüche asked himself: "They don't feel comfortable with German conditions and needs. But how should an American officer of perhaps 24 years of age be able to understand the German psyche at all?"

These passages reflect his frustration about his personal future, exacerbated by the material shortages and the cloak of mystery shrouding the Allies' plans. A lack of trust in the new governors was his answer. It goes without saying that Brüche's impatient demands about the infrastructure were totally unrealistic. But the dialog with the occupiers had at least been begun. In the coming months it became more productive elsewhere as well. In a diary entry at the end of June 1945 Brüche reported out of a "renovated institute that the war had left totally unscathed" in Erlangen about one of these dialogs. A former student of Pohl, Rudolf Hilsch (1903–1972), stoutly defended the demands of physicists as a group:

> Hilsch spoke for 4 hours long to two Englishmen and even if only 1% of it stuck, Saul must have turned into Paul. Lt. Comr. A. Elliott, RNVR, was the higher ranking of the two, who both listened with great interest to Hilsch's portrayals of the stance of physicists toward the party. Hilsch said what any other physicist would also have said. But whether just any physicist would have taken such pains with 2 Englishmen is very doubtful.

The Allies were in the difficult position of deciding whether to heed the call for immediate severe punishment of the guilty among the defeated enemy or to look to the future and foster their reeducation into good democrats.[66] Even clearly definable goals like dismissing the most seriously incriminated persons from university faculties conflicted with the equally clear goal of reopening important academic institutes as soon as possible. These included university clinics and hence physics institutes, where medical students received an important element of their training. As the period of occupation dragged on, the Allies, particularly the British and the Americans, were increasingly tempted to identify themselves with their charges, against the official prescription of maintaining "restrained arrogance" and at odds with the prohibition to fraternize.[67]

This same dilemma between the drive to purge and reform and practical considerations troubled the reopening of other scientific institutions and professional publications as well. The pressure to fall back on existing structures and staffing

---

[66]The Allied perspective on these conflicts is provided, e.g., by Balfour & Mair [1956] pp. 14–50, 162–183, Anonymous [1947] p. 243, and Kellermann [1978]. Tent [1982] goes into the guiding idea of "reeducation" as the Americans understood it.

[67]Officer Edward Y. Hartshorne (1912–1946) describes this tense situation in detail in his diaries and correspondence, published in Tent (ed.) [1998]. Hartshorne was responsible for the press and the reopening of the universities in the American sector. See there pp. 5f. on the prescribed attitude of "restrained arrogance" and the problems regarding the prohibition to fraternize. Ernst Brüche discusses the reaction when it was lifted, in diary no. IV, 14 July 1945 (BLM, box 7).

Fig. 8: American poster on reeducation, US zone around 1947. The fact that this poster (originally 84.9 × 62.4 cm in size, is in English indicates that the intended audience was members of the American Military Government. It also indirectly documents that in practice, the lack of ethical conduct was a problem necessitating such reminders. Source: Haus der Geschichte, Bonn, EB-no. 1994/04/0331.

continued to mount. The case of the Kaiser Wilhelm Society (*Kaiser-Wilhelm-Gesellschaft*) exemplifies this. This national research society was dissolved in 1945 only to be reopened under a new name: *Max-Planck-Gesellschaft*.[68] The Allies took their time approving a few of the traditional journals, such as the *Annalen der Physik* and the *Zeitschrift für Physik*. This left a vacuum during the first two postwar years that new publications were happy to fill: the *Zeitschrift für Naturforschung* and the *Physikalische Blätter*. The latter had originally been authorized by the chairman of the German Physical Society (*Deutsche Physikalische Gesellschaft*), Carl Ramsauer (1879–1955), in 1944 as the official publication of the "Information Office of

---

[68]See, e.g., Manfred Heinemann in Vierhaus & Vom Brocke (eds.) [1990], Hermann [1993] pp. 83ff. Cf. *PB* **2** (1946) p. 124, **3** (1947) p. 136. Oexle [1994], Lemmerich (ed.) [1998] p. 468, and Walker (ed.) [2003] pp. 40ff. discuss the background to Allied demands regarding the society and its renaming. Erich Regener's address in honor of Hahn's 75th birthday, in UAS, SN 16/26, p. 4, alludes to General Clay's initial view that the KWIs constituted a military threat. See also the transcript of a radio discussion between Hahn and Regener about the Max Planck Society, UAS, SN 16/5, esp. p. 3.

German Physicists."[69] Its first volume 1944/45 had unabashedly emphasized the crucial military importance of physical research. Its publishers nevertheless managed to obtain permission, first from the American and British sectors and soon afterwards from the French sector, to publish an "emergency journal," which appeared in obvious continuation of its former title as the *Neue Physikalische Blätter*. As we can gather from a carbon-copy circular letter by the founding editor to his colleagues from March 1946, Brüche envisioned "an undemanding journal to reestablish contacts within physics and to discuss issues of the day." So it quite specifically also set its sights on science policy:[70]

> Particularly our neighbor to the west seems to show an interest in the journal and in collaborating. I consider it beneficial, because I believe that in future, scientific relations with France, in particular, should be fostered on an equal basis.

With such headlines as "Research in distress!' the first issues of the *Physikalische Blätter* in 1944 certainly could have been interpreted as a political mobilizer of the country's last reserves. Isolated criticism was voiced about the journal's staffing continuity after the war, but it never became public:[71]

> Concerning Mr. Brüche, I do find it unfortunate that such a man as Mr. Brüche be the editor of the Physikalische Blätter. Judging from the content of the Physikalische Blätter, Mr. Brüche was, during the Nazi period, certainly moving in the Nazi wake. So I do wonder that this journal was given permission to appear again. But nothing can apparently be done about it practically, at the moment, but the Physikalische Blätter must remain the publication of the Physical Society.

This publication evidently also was subject to the scrutiny of Allied Control. Censorship is perhaps too strong a word, but it was clear that the good will at

---

[69]The background to the founding of the *PB* and its publishing outlet is discussed, e.g., by E. Brüche in *PB* **3** (1947) pp. 224–226 and Dreisigacker & Rechenberg [1994] in the commemorative issue; Rammer [2004] p. 299. See the documentation in BLM, esp. a statement for the Allied Control authority in box 5, folder 58 ("Erklärung über die Physikalischen Blätter"), dated 28 Apr. 1946. The 150-year-old *Annalen der Physik* was refused its bid for paper consignments: see the introductory article *PB* **2** (1946) p. 2. Other scientific journals are discussed in *PB* **2** (1946) pp. 111–114.

[70]Brüche's circular, March 1946: carbon copy in Walter Kossel's issue of the *PB*, kindly lent to me by Gerhard Rammer. On Franco-German relations see here footnotes 295 and 325 and Brüche's diary no. IV, end of Nov. 1945 (BLM, box 7).

[71]W. Meißner to M. von Laue, 15 Dec. 1947, AMPG, div. III, rep. 50, no. 25. Brüche elaborates on his motives in a letter to Max Planck, 21 Oct. 1945, ibid., no. 2390.

the outset could easily be spent. Self-imposed control and careful testing of the limits of free speech were the rules of engagement. Paper consignments played as important a role for the journal as its licensing. The first issues in 1946 were printed at a run of 5,000. Even that "did not even remotely suffice to cover the demand," but the run for early 1947 had to be reduced to 2,500 copies because only about 500 kg could be made available to them per quarter instead of over twice that amount. At the end of the second issue for 1947, Brüche wrote:[72]

> this had been interpreted by some readers as a sign of the Control authorities' purported discomfort with the journal. Pessimists were already saying that the journal would now go to its demise for economic reasons. The editor can declare today that the Military Government has restored the earlier printing and has moreover basically granted its approval for an increase in the printing by 50%.

Brüche took this upgrading to 7,500 copies as confirmation of his opinion that "honest discussion is regarded as detrimental only to those who are not able to distinguish between candid expressions of opinion and mean-spirited criticism; nor does the forwarding of contributions from America through the *Information Control Division*—the first among these to appear 'exclusive to your magazine' in one of the coming issues—speak against it."[73] This change of policy to include contributions by the ICD without comment approached the tolerance limit of the *Physikalische Blätter* in conforming with the new government. Tinctured stories, such as, "50 years of American physics" came dangerously close to the boundary between reporting and propaganda. Brüche laid particular store by independence.[74] The following issue (no. 5) already included a printed note on red paper, justifying a reduction from 40 to 32 pages owing to an acute paper shortage. It also urged its readers to send any available recyclable paper directly to the publishing house, which at that time was "Volk und Zeit," run by W. Beisel in Karlsruhe.

---

[72]E. Brüche: "Neue Auflage der Phys. Blätter," *PB* **3**, no. 2, p. 64 (closing date 28 Mar. 1947). Brüche mentions the amount of paper needed for the *PB* in his letter to R. Fraser, 11 June 1947, BLM, box 1, folder 2. Mannkopf refers to the administrative hurdles involved when paper consignments crossed zonal borders in his letter to Brüche, 26 June 1947, box 119, folder 308.

[73]Ibid. Balfour & Mair [1956] pp. 211–228 discuss the supervision of journals and other media in postwar Germany. On the ICD see also Tent [1982] pp. 77f., 86ff.

[74]Gordon F. Hull: "50 Jahre amerikanische Physik," *PB* **3** (1947), no. 4, pp. 107–110, among other notices of the same tenor. Brüche remembered his effort to maintain this independence in "Nec temere nec timide" in *PB* **6** (1950); draft manuscript AMPG, III, rep. 50, no. 370. See Dreisigacker & Rechenberg [1995] pp. F140f., Hoffmann & Stange [1997].

There was a clear political agenda attached to this hectic flurry of publications after 1945. Max von Laue spells it out in a letter to fellow theoretical physicist Fritz Bopp (1909–1987):[75]

> We Germans have lost our political position as a great power for the fore-seeable future. It is a matter of restoring our position as an intellectual great power and defending ourselves against stiff foreign competition, par-ticularly from America. Success here could possibly be a critical factor in the fate of the German nation, which stands the risk of being considered a quantité négligeable. Once established, history will too easily pass over us and move on to the order of the day.

This active-duty rhetoric was not new. After the First World War there were similar statements aimed at styling professional science as the nation's last bastion.[76]

(b) Covert reserve and distrust

Despite the official rhetoric, only a few physicists in Germany seemed to have viewed the arrival of Allied troops as a liberation from the National Socialist dicta-torship. Max Steenbeck (1904–1981), interned by Soviet soldiers more by chance than owing to his position as foreman at a Siemens factory, wrote retrospectively:[77]

> While we men sat in captivity in camps, our wives and children had to get by alone in a topsy-turvy world of violence, hate, and murder—and that was through our own making. No, on the eighth of May nineteen forty-five, I, in any case, absolutely did not feel liberated; perhaps liberated from the constant thought: Will you be alive tomorrow, at all? But that was insignificant against the life we then saw ahead of us.

Ernst Brüche's diary contains very similar thoughts, even though he was more fortunate. He was not rounded up while at work for a leading electric company:[78]

---

[75]Max von Laue to F. Bopp, 29 Jan. 1952, AMPG, div. III, Rep. 50, No. 311.

[76]For other examples of this rhetoric see, for instance, Heisenberg [1969] pp. 251f., Schüring [2006] p. 258.

[77]Steenbeck [1977] p. 151. Nor did the Allies have the impression that their presence among the Germans was seen as a liberation: see, e.g., Hartshorne in Tent (ed.) [1998] p. 19; Stone [1947a] p. 5: "not as liberators, but as conquerers." The bad conditions in some of these internment camps are described by von Salomon [1951] pp. 557–615. Fritzsch [1972] pp. 13f. cites a rounded total of 245,000 internees during the first months of the occupation; c. 100,000 of them were released again on 1 January 1947.

[78]E. Brüche, diary no. III (1 Apr. 1940–28 May 1945, entry on 19 Apr. 1945 (BLM, box 7). Cf. Schüring [2006] p. 238 for excerpts from the reminiscences of Ernst Telschow's secretary about the day after Allied troops marched into Göttingen.

"We sit there in the forecourt of the Krügel building [of AEG in Oberweissenbach] on a tree-trunk in the sun and feel like prisoners, which of course we are. We live under the most primitive conditions but even worse may well be in store for us. [...] Inside the factory no one wants to work anymore. It's so pointless. Do what, and what for?"

The arrival of Allied troops in a town was soon followed by the command to shut down the local university or college. A directive of the *Joint Chiefs of Staff* (JCS 1067), initially only released to a few people in leadership positions on 21 May 1945, later to be published on 17 October 1945, determined that only "after removal of the characteristic traces of Nazism and of Nazi staff" could a reopening of educational institutions be permitted.[79] The Soviet Military Administration of Germany (SMAD), officially constituted on 9 June 1945, passed corresponding laws and edicts in the Soviet zone. Directive No. 40 of 25 August 1945, On the Preparation of Schools for Instruction and No. 50 of 4 September 1945, On the Preparation of Universities for the Commencement of Courses and Control of Their Activities, marked the beginning of denazification among teachers; Directive No. 35 of 26 February 1948 marked its official end.[80] Although this purge in the Soviet occupation zone was initially carried out with greater effectiveness, actual implementation of the central directives fluctuated widely from region to region. As Feige points out, "Scientists in Leipzig and Dresden who had been dismissed as a consequence of their membership in the NSDAP had no trouble finding new positions in Rostock, Halle or Jena." At the Mining Academy of Freiburg in Saxony, for instance, 60 members of the staff were dismissed on 9 November 1945 because of their Nazi affiliations (12 of them were professors). Nevertheless, a month later the majority of them were in fact asked to resume their duties, which just happened to involve research for the Soviet military services.[81] In wise premonition, the University of Leipzig categorized all its political offenders by the middle of 1945 into: "indispensable," "indispensable until further notice," or "dispensable."

[79]See, e.g., Balfour & Mair [1956] pp. 23ff., 66f., Treue [1967] S. 22-24, Latour & Vogelsang [1973] pp. 17, 177, Hauser (ed.) [1987] pp. 73f. Schlüpmann [2002] p. 454 points out that this directive was replaced by JCS 1779 on 11 July 1947, emphasizing reconstruction and the changed course. For the response by physicists to these directives, see e.g., Brüche's columns on a "Revision of prejudice," *PB* **5** (1949) p. 48, and "I certainly do hear the message," ibid., p. 392.

[80]The uneven implementation of SMAD university policy is discussed by Heinemann [1990] part 4 and Feige [1992]; see there p. 1172 for the following quote.

[81]See Haritonow [1992] and on the following, Welsh [1985] p. 345, Feige [1994].

The dangling Damocles sword of dismissal for political involvement was the more threatening since for a long time no one knew how exactly such "traces of Nazism" were to be defined: on the basis of formal criteria like membership in National Socialist organizations or merely from eye-witness reports and affidavits? Some Allied Control officers tried not to rely on the mere formality of membership in the National Socialist Party or its suborganizations, looking also at individual political affinity with such Nazi ideologies as Aryan superiority, the authoritarian *Führer* principle, or the exercise of unlimited dictatorial powers.[82] A "purging commission" autonomously established by the University of Tübingen in May 1945 worked with three categories: (I) "aggressive National Socialists" (mainly involving the figureheads of fields specially promoted by the Nazi state), (II) highly placed "convinced" National Socialists who had distanced themselves perceptively from the "excesses" of the Nazi regime, and (III) nominal party members guided by "idealistic motivations."[83] According to the formal criterion of party membership, all of the above were political offenders. This finer division into three grades had the obvious goal of limiting immediate dismissals and suspensions to the gravest offenders of category (I).[84] The thus restricted circle of intolerable individuals comprised:

- Those whose "exercise of office or public statements [had] seriously damaged the reputation of the university and the esteem of the academic field through compliance with National Socialism."

- "Informants and spies. Their violations of the basic rules of human decency and of the corporate spirit and collegiality are by no means excusable by having been committed on official orders."

[82]See, e.g., Balfour & Mair [1956] p. 52. Hartshorne's classification of political attitudes held by university professors are described by Tent (ed.) [1998] p. 103, who also reported on the frequent distinction made between party comrades and "only nominal Nazis" (*Parteigenossen* (PGs) versus *Karteigenossen*), ibid., p. 75. See also the quote here on p. 117.

[83]Wischnath [1998] pp. 107f. discusses the categories of the Tübingen "commission d'épuration." The policy in the French occupation zone is treated by Henke [1980], Heinemann [1990] part 3, Fassnacht [2000] pp. 110ff. The concept of "auto-purification" (*auto-épuration*) by the General Administrator of the French Zone, Emile Laffon, is discussed more generally in Seemann [2002] pp. 39ff., 73ff., 85ff., 211ff., its four-stage demise on pp. 134ff.

[84]See Wischnath [1998] pp. 108f. for the exact figures at Tübingen: 10 of 11 chair-holders in category I were discharged, but only two in category II and none at all in III. Similar categories and dismissal figures at Freiburg are available in Seemann [2002] pp. 85ff., 195f. A comparison with other occupation zones is made in Söllner (ed.) [1986] pp. 185ff.

- "Representatives of fields that were newly created by National Socialism for political reasons without substantive legitimacy."

- "Those who attained their positions without sufficient academic qualifications owing to a National Socialist mentality."[85]

There was more behind this restrictive structure than purely pragmatic concerns about keeping educational institutions running, despite a lack of suitable staff replacements. Rectors, deans, and other university leaders—indeed even members on denazification or purging boards—often acted on a skewed sense of solidarity toward their incriminated fellow colleagues. When this strategy of limiting the sanctions and punishments to a few extreme cases seemed to be taking effect, while elsewhere rigorous waves of dismissals were being carried out, the formerly disadvantaged were understandably outraged, and a general sense of insecurity spread about the way justice was being meted out. Ambiguous, at times even contradictory directives gave local occupation administrators and officials broad discretionary powers that varied considerably among the different zones and even among the different university locations. This only strengthened the impression of unfair treatment.[86] Many implicated persons also tried to use this situation to their advantage to shape better initial conditions for a new start in life. That might even entail adeptly placing obstacles in the way of personal competitors and opponents.

A new wave of accusations and anonymous denunciations similar to the early period of the Nazi regime swept over the land. Some of the charges targeted real culprits, while others, it turned out, were just a settling of old scores. The postwar rector of the Friedrich-Wilhelms-Universität in Bonn, Heinrich Matthias Konen[87] (1874–1948), thought it necessary to urge students in his welcoming address to the new enrollees at the end of 1945 not to compromise the "national dignity" and "to address as few letters to the military government as possible," in other words, to refrain from submitting any denunciations. Geoffrey Bird, the British

---

[85]Thus the list of criteria of one "commission d'épuration" in Freiburg. Memorandum dated 6 July 1945, cited from Fassnacht [2000] pp. 116f.

[86]See Fritzsch [1972] p. 18. Sommerfeld's character reference for Heinrich Ott from 13 Feb. 1946 is one example of a physicist complaining about unequal treatment: http://www.lrz-muenchen/~Sommerfeld/gif100/05160_01.gif

[87]The spectroscopist Konen was a renowned politician of the Center party who had been forced into early retirement in 1933 but presided over the University of Bonn's reconstruction after the war: see Gerlach [1980]. His unpublished address of 1 Dec. 1945 is discussed in Wolgast [2001] pp. 292, 297.

University Education Control Officer in charge of preparing the reopening and supervision of the University of Göttingen, recalled:[88]

> The task of finding reliable members of the staff was made no easier by the considerable amount of conflicting evidence and the denunciations which were forthcoming from a few professors about some of their colleagues. Sometimes these denunciations proved correct, while a few, less reliable, were inspired by personal jealousies and grudges, and it was not always easy to elicit the facts.

Fig. 9: Excerpt from an appeal by the Ministry for Political Liberation to collaborate in finding the guilty, 1946. Source: Stadtarchiv Karlsruhe, also online at www.hbg.ka.bw.schule.de/publikat/ka45

---

[88]Bird [1978] p. 147. The DPG expelled Steenbeck from the society's board in December 1945. Ramsauer justified this on Steenbeck's party membership and that he was in Russian hands: Ramsauer's note in the file 18 Dec. 1945, DPGA.

The first years after 1945 at the University of Tübingen were also evaluated as follows:[89]

> Conflicts, denunciation and mutual distrust fragmented the academic community of the postwar period, although from the outside its defensive opposition to the new government and the occupying force sometimes made it appear like a united bastion. [...] Those who had a name in science and did not diverge in their political mentality and habitus from mainstream academia, i.e., the professorial educated class who were socialized as conservative nationalists, could count most on corporative protection.

John Gimbel, himself an officer of the occupying force as a young man, wrote in his book *Science, Technology and Reparations*: "The Germans continued to be *fearful and suspicious*, and eventually their worst fears came to pass."—He was referring specifically to the abuse of Allied Control laws for industrial espionage.[90] Allied Control Commission (ACC) Law No. 29 stipulated that "any of the four powers in occupation of Germany ... may request in writing an authenticated copy of any book, paper, statement, record, account, writing or other document from the files of any German industrial, business, or commercial enterprise." By 30 June 1946, various branches of the US *Field Information Agency, Technical* (FIAT) had conducted over 3,400 trips within Germany to investigate so-called "targets" of interest. It had processed over 29,000 reports and confiscated 55 tons of documents and sent them over the Atlantic to the US. Once there, they were made generally available to whoever took an interest in them.[91] In principle, experimental research in nuclear physics was categorically forbidden in defeated Germany. An exception was made for research exclusively for medical purposes and savvy scientists grasped this opportunity to sidestep the regulation. Wolfgang Paul (1913–1993) in Göttingen only had to redefine his nuclear experiments with an electron accelerator accordingly.[92] But Law No. 22 On the Surveillance

---

[89] Palatschek in: vom Bruch & Kaderas (eds.) [2002] p. 405. Wischnath [1998] offers the particular perspective of that university's last Nazi rector, Otto Stickl. A "persistence of a denunciation habit" is also mentioned in Tent (ed.) [1998] p. 86.

[90] Gimbel [1990] p. 180 (emphasis mine), particularly concerning the USA. For criticism of Gimbel's theses see Volker Berghahn in Judt & Ciesla (eds.) [1996]; regarding the British see Farquharson [1997]. Wolfgang Finkelnburg wrote to Arnold Sommerfeld on 28 May 1947 about "bitter feelings, that always need a long time to die out after a war." http://www.lrz-muenchen/~Sommerfeld/KurzFass/04690.html

[91] See Matthias Judt in Judt & Ciesla (eds.) [1996] p. 32 and Werner Abelshauser, ibid., p. 111 on the British 'Operation Plunder.'

[92] See Friess & Steiner (eds.) [1995] pp. 269f., 201 (excerpt from an interview with Paul, who names Ronald Fraser as instrumental in obtaining this coup); see also the elaborations in Rammer [2004].

of Materials, Installations, and Equipment in the Area of Atomic Energy and Law No. 24 On the Surveillance of Specific Objects, Products, Facilities and Apparatus issued by the *Allied High Commission* (AHC), which came into force as late as April 1950, placed such cramping limitations on German nuclear research that Heisenberg as chair of the German Research Council (DFR) spoke of a "strangulation" of this branch of research.[93] Four years had to elapse before there was any movement toward lifting the ban on war-related research. The protocol on the termination of the occupying government in the Federal Republic of Germany was signed in Paris only in October 1954 and ratified in 1955. However, the Aerodynamic Testing Station at Göttingen, which had been stripped almost completely bare of its installations in 1945, had to wait until the 1960s to return to full operation with state-of-the-art apparatus.[94]

Profound distrust is thus a general characteristic of the postwar mentality of scientists and engineers. The following notice in the *Physikalische Blätter* exemplifies this climate of suspicion and the resulting misrepresentations and biased self-portrayals of physicists in Germany before the Allies during the early period:[95]

> Whoever went to an American authority with some petition regarding the field of science during the first months of the occupation did well not to use the words "science" and "research" initially, confining the argument to aspects of science education. The American press and perhaps also

Harteck's assistant Wilhelm Groth was likewise able to continue his work on uranium enrichment with gas centrifuges as early as 1946, thanks to the tolerant approach of the British control officers, as Walter Kaiser reports in König et al. (eds.) [1997] p. 250. Cf. also the article "Germans woo atom lore" in the *NYT* of 13 Dec. 1949, where Heisenberg, referred to as "Germany's top atom smasher" and "Bonn regime physicist," is reported to have said: "German physicists, now working with Allied permission on medical aspects of atomic research, wished to extend their studies to other atomic fields."

[93] See the article "Control of Nuclear Research in Germany," intended for publication in the *Bulletin of the Atomic Scientists* in July 1950 together with a commentary by W. Heisenberg in the form of a letter of protest to Mr. K. H. Lauder, Director of the Research Branch Göttingen. Although page proofs still exist among the Bulletin's files, the article itself never actually appeared: RLUC, box 32, folder 7. Compare the contributions by Regierungsdirektor Friedrich Frowein, head of the German research supervisory office in Hessen, on research control in West Germany: *PB* **6** (1950) pp. 222–225 as well as "Law 22 yet again" ("Nochmals Gesetz 22") ibid., pp. 316–319 with critical commentary. On the German Research Council see also Stamm [1981], Eckert [1990], Carson & Gubser [2002].

[94] See Ciesla in Judt & Ciesla (eds.) [1996] on the advanced position of German wind-tunnel technology until 1945 and on the postwar transfer of this knowledge (along with some of its specialists) to the USA. The American Air Force (Wright Field, Ohio) and the aircraft industry were the main beneficiaries.

[95] [E. Brüche]: "Förderung der Forschung," *PB* **3** (1947), issue no. 12, p. 432.

the exaggerations of the National Socialist Ministry of Propaganda had fed suspicions that German laboratories were dangerous witches' kitchens needing to be forbidden, one and all, upon mere suspicion. Maybe the points of the Morgenthau Plan that, as an agrarian country and exporter of raw materials, Germany had no more use for research also had something to do with it.

Interest in taking public office was also very low, because many potential candidates were afraid of being held responsible for the vexations with the occupying forces and being stigmatized as a collaborator. Max von Laue told his son about a conversation with the governing mayor of Hamburg:[96]

> He said he was already over 70 years old and had little inclination to assume a state office again now. But he said that anyone doing so now in Germany had to count on soon being singed by it, at least politically singed. And that was why younger people were right if they didn't come in droves seeking it. It didn't matter so much for old folks. So that was why he had taken the burden upon himself. I mention this because one hears this idea frequently aired.

Distrust on all sides remained the order of the day in subsequent years as well. At the end of 1947 von Laue complained to an English colleague about the difficulties they were having filling a director's position at the German bureau of standards, the *Physikalisch-Technische Reichsanstalt* (PTR), and about the high turnover rates in the civil service:[97]

> This bad situation cannot be done away with at the moment. For that, the whole situation would need to calm down considerably, which could only come from the highest political levels. Peace has to come, and not just peace on paper, but peace of mind, whereby the terrible distrust, everyone against everyone, which poisons all human relations nowadays, finally stops. But it surpasses the power of scientists to bring this about.

Klaus Clusius (1903–1963), director of the Institute of Physical Chemistry and for a while also dean of the University of Munich, complained soon after its reopening that it had taken place "in due pomp and ceremony with the dismissal of most of its professors and assistants." His other letters are full of cynicism, bitterness, and fear of worse to come. He complained, for example, that Heisenberg's

---

[96] Max to Theo von Laue, AMPG, div. III, rep. 50, suppl. 7/7, sheet 4 (undated, c. Feb. 1946).
[97] M. v. Laue to Charles Darwin, National Physical Laboratory, 12 Dec. 1947; DPGA no. 40046; furthermore Zeitz [2006] pp. 126-151 on von Laue's role concerning the refounding of the PTR.

appointment would never work out for purely political reasons, because "at the behest of men of this terrestrial world" he has to stay in the British sector for the time being:[98]

> Measured against the enormities that we have the dubious pleasure of bearing witness to, this issue seems a triviality. But set yourself back in time to 100 years ago and imagine what would have happened if that which Europe is being blessed with today had taken place then, so ... we better keep quiet!

The British zone maintained a comparatively liberal attitude toward university affairs.[99] But not everything was tolerated there either. The nuclear and atomic physicists released from internment at Farm Hall were told what they were allowed to say in public and they were initially only permitted to reside within the British zone. Most of them went to Göttingen, where they flocked around Heisenberg in the relocated Kaiser Wilhelm Institute of Physics in a building made available to them by the British on the premises of the former Aerodynamic Testing Station. Carl Friedrich von Weizsäcker (born 1912) reported at the end of August 1946:[100]

> Immediately after our return to Germany we wanted to publish a brief report on our research [on the uranium machine]. This was prohibited at the time by the English. I have not been able to establish whether today we are able to publish such a report.

Their request became pressing after Samuel Goudsmit published in numerous articles and later in his book *Also* a quite one-sided version of the reasons for the failure of the German uranium project. According to Goudsmit, none of the collaborators ever realized that the true solution lay "far beyond the reach of an academic laboratory" involving "enormous industrial resources." Heisenberg, he contended, had worked in a vacuum "whose only atmosphere constituted what he

---

[98]K. Clusius to K. F. Bonhoeffer, 11 Dec. 1945 or 21 Jan. 1946, AMPG, div. III, 23, no. 14.1. Interestingly, Officer Edward Y. Hartshorne did not think much of Clusius's character: "Dean Clusius of the Naturwissenschaftliche Fakultät. [...] He made an ill-informed and irresponsible impression." (Tent (ed.) [1998] p. 289.)

[99]Comparative insights are provided by the contributions to Heinemann (ed.) [1981], [1990], and Axel Schild: "Im Kern gesund? Die deutschen Hochschulen 1945" in König et al. (eds.) [1997] pp. 228ff. The British zone is discussed in Heinemann [1990] part 1, Bird [1978], Phillips (ed.) [1983], Birke [1984], and Foschepoth & Steininger (eds.) [1985]. Parker [1987] outlines the general change in attitude by the British toward Germany around 1946.

[100]C. F. von Weizsäcker to E. Brüche, 27 Aug. 1946, Brüche papers, Mannheim (BLM).

was capable of producing based on his own expertise."[101] In 1948 Heisenberg was at last granted permission to publish detailed reports about the German "large-scale preparatory tests for the construction of a uranium machine" for the *FIAT Reviews* and elsewhere, to counter such misrepresentations.[102] This episode reveals the unfavorable effect of Allied publication restrictions on general assessments of the situation right after the war. German physicists chaffed impatiently against the unaccustomed bit.

A letter to the experimental physicist Walther Gerlach (1889–1979), until 1945 president of the physics section of the Reich Research Council and successor to Abraham Esau as Reichsmarschall of Nuclear Physics, is indicative. Its author, the Stuttgart metallographer Werner Köster (1896–1989), claimed he felt less free to express his opinion openly in 1947 than *before* 1945:[103]

> Dear Gerlach!
>
> Your straightforward statements delighted me. It was as if you were standing before me, the way you used to do, soundly scolding with eyes ablaze. I felt the refreshing breeze, the carelessness that I sorely miss among those around me here. People are all so timid, perhaps justifiably so, for one can't speak one's mind as freely now anymore as during the Third Reich. I had to endure a denunciation in 1945 already on a trivial matter. The mayor of Urach, a Social Democrat, told me at the time: Herr Professor, today you have to be even more careful about what you say and remark than under Adolf Hitler. You notice that everywhere.
>
> And that is a major reason why I decided to ask you for something after all. Nowadays one has no contacts with influential people. Nobody can evaluate the accomplishments, the people, the value of an institution. Cleaning women and dubious people straight out of concentration camps decide over the weal and woe of persons unknown to them. We recently had to stave off an attack here from a man of the latter sort, who had submitted an evaluation to the trial court [on] someone he hadn't ever even met. This opinion had been simply invented. That is why it might be good if you would write a letter to the Minister of Culture ... in which you convey to him roughly the following.

At the Institute for Metal Research of the Stuttgart Polytechnic the atmosphere was very tense. The possibility of it being closed down altogether loomed large.

---

[101] Goudsmit [1946] thus paraphrased by Brüche in an enclosure to his letter to C. F. von Weizsäcker, 20 Aug. 1946. A German translation appeared that year in the *Physikalische Blätter*.

[102] See Heisenberg & Wirtz [1948], Heisenberg [1949]; cf., e.g., Walker [1990] [1993] for a survey.

[103] W. Köster to W. Gerlach, 20 Feb. 1947, DMM, Gerlach Papers (on obsolete letterhead of the *Zeitschrift für Metallkunde* (NSBDT and VDI in NSBDT), quoted in Schlüpmann [2002] pp. 398f.

Anyone familiar with such letters soliciting personal references, or whitewash certificates (*Persilscheine*), will immediately sense more concrete fears behind such reservations about the new decision-makers. As institute director and professor of metallography since 1934, Köster was seriously worried about a verdict by the denazification tribunal:[104]

> Glocker and I have been suspended from our duties for 5 months. The precondition for our return, which ultimately depends on the Americans, is denazification. Before that happens we can't move an inch. So German voices of authority have to procure the basis for our clearance. They should help their own people decide in Germany's favor.

Köster had been a member of the National Socialist Party since 1940, but his leading positions in the Nazi League of German Technicians (NSBDT) and representing metal research on the Reich Research Council were harder to wave aside. After another page describing in-house matters at the institute, Köster continued:[105]

> And thus we come to personal matters, as depressing as they are unavoidable. A man concerned with such things recently said to me, the whole thing is a business matter to be negotiated. Before an immoral law, all decent impulses must remain silent. So you are going to have to say a few words about my character: whether or not I am mature enough for the new Germany, how my direction at the institute was, whether we were rabid Nazis, and so on. I don't want any of the usual exonerating statements from you, which I so thoroughly hate, but a clarification to the men now in government. How are they supposed to know? [...] Today we must defend ourselves. Today we are the suspects.

This crass assessment of the situation was not just an isolated instance. There is a similar statement in Ernst Brüche's diary, stemming from worries about the fate of the AEG factory in Nuremberg, which "might possibly become leaderless from the denazification":[106]

---

[104] Ibid. Anyone required to appear before the tribunal was temporarily removed from his or her position, being only permitted to work in a "subordinate" position pending the verdict.

[105] Ibid. On metal research during the Third Reich, particularly in Stuttgart, see Maier [2002]. Köster was interned on the charge of having contributed decisively to the German war machine. This included torpedo design, zinc ignitors, and sensitive magnetic-field detectors used in mines. In 1947, Köster was classified as a fellow traveler, fined, and allowed to continue his work from early 1938 on: see Helmut Maier in Becker & Quarthal (eds.) [2004] pp. 178–181, Grüttner [2004] p. 96.

[106] E. Brüche: diary IV, 11 Oct. 1945 (BLM, box 7).

These mindless dismissals of all former Nazis could drive one to desperation. The method only shows that the Americans are no smarter than their
predecessors, the Nazis. What did a reasonable man say to me yesterday?
From a mild dictatorship with its faults we have now arrived at a severe
dictatorship.

A questionnaire from 1946 in the American sector reveals how direly denazification
was needed among the population. Given the alternatives: Nazism was (i) a bad
thing, (ii) a good thing, or (iii) a good thing badly carried out, 40 percent of the
respondents chose the third. When the questionnaire was repeated in 1948, this
figure actually rose to 55.5 percent.[107] At the end of 1951, 35 percent of West
Germans still had a better impression of National Socialism in retrospect than in
1945, whereas only 16 percent found it worse.[108]

(c) Stubborn resistance to Allied Control

We have just seen a surprising willingness by two physicists in responsible positions to bend history for the sake of an institute's future in the "new Germany."
In the manner of social constructivism *avant la lettre*, denazification had become
negotiable. "Decent impulses" (evidently meaning pangs of conscience about
withholding incriminating information concerning Köster's past) must be sup-
pressed before an "immoral law"—out of a sense of duty. Allied Control directive
nos. 24 (12 January 1946) and 38 (September 1946) specified the removal of all active members of the National Socialist Party and other persons opposed to Allied
objectives from public and semi-public offices and from responsible positions in
major private companies.

Gerlach issued whitewash certificates not just for his close friend Köster, but
also for a former leading science policy-maker, Rudolf Mentzel (1900–1987). This
SS regiment leader had been president of the German Research Association (DFG)
and head of the Science Office in the Reich Ministry of Education. Before 1945,
Mentzel had certainly not always lent his support to Gerlach as head of the physics
sections of the DFG and the Reich Research Council (RFR), so this solidarity
is surprising.[109] Ernst Telschow (1889–1988), the long-time general secretary

---

[107]Balfour & Mair [1956] p. 58; cf. ibid., pp. 169–183, 331–334, Merritt & Merritt (eds.) [1970]
pp. 30ff., 105, 162f., 171, 210, 295, Tent [1982] pp. 83–109, Kielmansegg [1989] pp. 20f., 31ff.,
Vollnhals (ed.) [1991], and the following text, esp. footnotes 125, 133 on denazification.

[108]See Merritt & Merritt (eds.) [1980] pp. 7, 151, based on answers by approx. 1,200 W. Germans.

[109]See DMM, Gerlach papers, file on the denazification process, in English trans. in Hentschel
(ed.) & Hentschel (ed. asst./trans.) [1996] pp. 403–406. For the contextual reasons and other

of the Kaiser Wilhelm Society, also received exonerating personal references, even though he was known for implementing the policy of expulsion from 1937 onwards. Exiles who remembered only too well his conduct during the Nazi regime were enraged to find out that even Otto Hahn and Max Planck (1858–1947) had signed such references.[110] A former member of the Security Service (SD), who served as a witness at Mentzel's tribunal in Bielefeld, complained bitterly in a letter about[111]

> the obvious lack of instinct and courage of conviction among many scientists [...]. Mentzel was able to present dozens of personal evaluations by eminent scientists using soft and kindly words of no factual import about the activities of the once so notorious and feared Mentzel. Many important incriminating points that should have been cleared up, if only in the interest of the reputation of science, were simply swept under the carpet. Professors Kuhn and Butenandt, who had been forced to sign the awful letters declining their Nobel prize in Mentzel's office, described this event in a way that put Mentzel in a positive light. [...] Mentzel's part in the experiments carried out in concentration camps, which so severely damaged the reputation of German science, also disappeared under the carpet. In reply to questioning by the court Mentzel declared he knew nothing about it, and that was that. (Dr. Sievers, who had organized these experiments—and was hanged in the Nuremberg tribunal of medical doctors—had been named by Mentzel as his deputy in managing the Reich Research Council!) In reconstructing Debye's case, Dr. Telschow, who is currently still general director of the Max Planck Society and was closely allied with Mentzel, came forward in an attempt to clear Mentzel, etc.

positive and negative statements about Mentzel, see Schlüpmann [2002] pp. 417f. and footnote 124 below as well as p. 57. M. von Laue wrote O. Hahn, 23 Aug. 1946, about Gerlach's political attitude: "during the war years I also heard how his stentorian voice resounded through the KWI of Physics in Dahlem: 'We have to be victorious!,' a sign that he wished for Hitler's victory and continued rule." AMPG, div. III, 14A, no. 2462. But von Laue's protest against Gerlach's involvement in the reopening of the physical society was ignored and on 11 Sep. 1946 Gerlach became one of the founding members of the MPG in the British zone: see Heinrich & Bachmann [1989] p. 187 and Heinemann in Vierhaus & Vom Brocke (eds.) [1990].

[110] See Schüring in vom Bruch & Kaderas (eds.) [2002] p. 456. Sime [2006] p. 37, unaware of the typical elements of this mentality among physicists, exaggeratedly personalizes Hahn's tendency to turn into a common cause the troubles of those whose political views he had previously despised.

[111] Helmut J. Fischer to E. Brüche, 3 July 1949, BLM, box 3, folder 42. Fischer joined the Nazi party in 1933 and worked part time from 1936 for the SD; full time 1938–44. He joined the SS in 1938 and was promoted to captain in 1941. In 1944/45 Fischer served as science rapporteur for the Reich Central Security Office (RSHA). See Grüttner [2004] p. 49 and Fischer's memoirs *Erinnerungen*, published in 1984/85.

Ernst Brüche defended former high-ranking members of the Security Service and leaders of the Reich Youth as well as a censor at the Ministry of Propaganda. He was grateful for their help in procuring paper for the *Physikalische Blätter* and publication permits and for helping obtain recalls of physicists from the Army. Brüche vaguely acknowledged their efforts in support of science and fundamental research.[112] As one defendant suggested to his supporting witness, this all happened in a "war against one-sided applied research" and in "open criticism of prevailing conditions, party corruption, ignorance, unprofessionality, the resource egoism of leading party names, etc.," but decidedly not "in order to extend the war, only out of a true understanding of the ideals of science." Nevertheless everyone must have been aware that these actions had a stabilizing effect on the National Socialist system. Pascual Jordan (1902–1980) also received an exonerating certificate from Heisenberg, Pauli, and others, despite active membership in the party and the Nazi Storm Detachments (SA), and despite numerous publications heavily laced with Nazi rhetoric.[113] Heisenberg argued, for instance, that he never seriously believed that Jordan was a genuine National Socialist and left unmentioned Jordan's repeated advocacy of the "will for might," the *Führer* principle, and the development of weapons technology. The Soviets also sought a character evaluation concerning Jordan's political past when they were considering appointing him director of the biophysical research complex in Buch near Berlin in 1947/48. Robert Rompe (1905–1993) stylized his wartime friend as a victim of his bourgeois family background and the fascist milieu affecting his actions at the University of Rostock and in the *Luftwaffe*. Despite these bad influences, he argued, Jordan's Nazi affiliations had just been a formality. Josef Naas (1906–1993) did concede an ideological fault in Jordan's character but hoped that it could be corrected in the good company of Rompe and other cadres of the ruling East German Socialist Party. So despite the anti-fascist rhetoric, even in the Soviet occupation zone, there was a willingness to look the other way when the persons in question were indispensable for the establishment of new research institutions.

---

[112] See E. Brüche's statement on the tribunal case of Helmut Fischer, 2 Oct. 1947, BLM, box 1, folder 2, and Helmut J. Fischer to E. Brüche, Sep. 1947 with a two-page addendum of exonerating points (quoted in the following), and Brüche's exonerating certificate for the censor Ludwig Bücking and the main district leader of the Reich Youth, Heinrich Hartmann, ibid., box 90, folder 225 as well as for the former *Reichsjugend* leader Artur Axmann in box 104, folder 261.

[113] Jordan's publications from the Nazi era are discussed by Wise in Renneberg & Walker (eds.) [1994] pp. 247ff., Beyler [1994] pp. 224ff., 465ff. See also here p. 107. On Jordan's denazification see Wise in Renneberg & Walker (eds.) [1994] pp. 251f.; Beyler [1994] pp. 474f., 478f., quotes from Rompe's and Naas's evaluations of Jordan's political attitude, dated 4 Oct. 1948.

Max Born, by contrast, declined a request by his former assistant and close coworker. Jordan had sent him a typical letter in 1948 underplaying his political activities, emphasizing his part in the fight against the 'Aryan physics' movement and his decision not to participate in the German rocket or nuclear programs. Instead of getting the hoped-for affidavit, he received from Edinburgh, where Born was living in exile, a list of the relatives and friends Born had lost in the Holocaust.[114] In the western zones Arnold Sommerfeld contributed to the whitewashing with at least 12 documentable exonerating affidavits against only two refusals. He declined Hans Kneser (1901–1985) because too much time had elapsed since their acquaintance. The former dean, Karl Beurlen (1901–1985), was the only applicant he refused outright for political reasons.[115] People with particularly shaky credentials tended to collect as many of such statements as they could so that they could pick out the best ones for their hearings. Gerhard Rammer has managed to find 57 whitewash certificates for Karl-Heinz Hellwege (1910–1999). This ranking SA officer had been dismissed in 1946 on various other charges as well. His certificates, some in multiple copies, had been signed by colleagues, relatives, acquaintances, and even his landlord.[116] In a confidential letter from 1947 Max von Laue described the guiding principle behind such positive characterizations by him and his colleagues: "We are trying to implement a policy here in physics that is unfortunately not being adopted by the state, namely, carrying out one big amnesty for all Nazi fellow travelers, after harshly condemning the real culprits."[117] Carl F. von Weizsäcker adopted a similar tone

[114]P. Jordan to M. Born, 23 July 1948 and 15 Aug. 1948, SBPK, Born papers, 353, cited by N. Wise in Renneberg & Walker (eds.) [1994] p. 252, also reprinted in facsimile in Hoffmann [2003]. After 1945 Pauli also seems to have assessed Jordan's political milieu as a mitigating circumstance: see Schücking [1999] p. 26: "Richard Kuhn had no excuse for having been a Nazi. But Herr Jordan had. He was a professor at Rostock! (Rostock was considered the Outer Mongolia of German universities). Jordan winced."

[115]On Sommerfeld see the website of the Sommerfeld project under 1946 and 1947. Sachse [2002] p. 231 speaks, with reference to Walker, of as many as some 60 *Persilscheinen* issued just by Sommerfeld and Heisenberg. On Hahn see Sime [2006] and Walker [2006]. According to Schüring [2006] pp. 278ff., Hahn refused to write an affidavit for the former general director of the KWG, Friedrich Glum.

[116]See Rammer [2003] pp. 94f. and [2004] pp. 102–107 for interesting excerpts from this letter. It indicates a significant range of what evaluators were able to certify about this convinced National Socialist physicist: some colleagues distanced themselves between the lines: "His specialty was relatively far away from ours, likewise regarding the military projects he was working on" or "I ... never had reservations about telling him my opinion frankly, even when it did not agree with his." Others constructed a picture of Hellwege as an opponent of National Socialism, referred to his religious connections, or emphasized his personal merits.

[117]M. von Laue to H. Pechel, 11 Nov. 1947, DPGA No. 40048. M. von Laue wrote to his son

in a letter to the editor of the *Physikalische Blätter*, arguing against expanding the numbers of the accused:[118] "I also think that persons who have not been drawn into the debate about past errors should not, on principle, be included now wherever it can possibly be avoided." This was not just out of pity for those under fire. It was following a maxim of damage limitation for the discipline. Physics as a whole could only suffer by deeper delving into the past. By this time Otto Hahn also shared this attitude. "Drawing the line" ("Schlußstrich-Mentalität") is how Norbert Frei described it, for the sake of the young Federal Republic of Germany. When a doctoral student confronted Hahn with incriminating facts about Pascual Jordan and Herbert A. Stuart (1899–1974), he replied to the young woman:[119]

> In such cases I am often asked how I should respond. If it doesn't involve blatant cases, I answer that I certainly won't do anything for the gentlemen, on the other hand, I won't actively bring charges against them either. I am reluctant to continue to add to all these unpleasant things. We had enough trouble with all that snooping and telling off during the Third Reich, and I don't think that after these gentlemen have their tribunal proceedings behind them and are relieved about it, that they will suddenly come forward again as active or potential Nazis.

Even among emigrés there was opposition to the Allied denazification scheme. Lise Meitner, for instance, generously issued certificates to people whom she knew had once actually denounced her politically.[120]

Theo von Laue along similar lines on 16 July 1946, AMPG, div. III, rep. 50, suppl. 7/7, sheets 27f.

[118]C. F. von Weizsäcker to E. Brüche, 10 April 1947, BLM, box 119, folder 308. The person in question was Johannes R. Malsch (1902–1956): "I would be very sorry if more serious difficulties than he already has were to arise out of the information about his participation in the Munich meeting. [...] It is a matter here of human tact." Self-interest on Weizsäcker's part must have played a role besides the arguments he layed out. He was reluctant to become directly involved in such debates.

[119]O. Hahn to U. Martius, 12 Nov. 1947, AMPG, div. II, rep. 14A, no. 2726 (G. Rammer kindly brought this to my attention). For excerpts from this Jordan correspondence before 1945, see Hoffmann [2003]; cf. Schücking [1999] p. 28 for the following postwar dialog between Jordan and Pauli about these passages: " 'Herr Jordan, how could you write such a thing?' To which Jordan retorted: 'Herr Pauli, how could you read such a thing?' " This insinuation that writing texts bristling with Nazi jargon was on a par with reading them obviously did not go down well with Pauli. But it demonstrates an absence of any real feelings of guilt, or at very least, the front shown in public: both were equally detrimental to any honest discussion.

[120]L. Meitner to J. Franck, 10 July 1947, Franck papers, RLUC, box 5, folder 5. For excerpts from letters by Hermann Fahlenbrach and for replies by Droste and Meitner, see Sime [1996] p. 350, Schüring [2006] pp. 280f. Hahn's whitewash certificate and another letter to Droste from 31 Jan. 1947 of a quite different tone are mentioned in Sime [2006] p. 34. Heisenberg chose the following

Fig. 10: Poster by the *Freie Demokratische Partei* (FDP, the liberal democratic party, which at that time was catering to the conservative right) for the first general elections in 1949. The perceived injustices of denazification was a supporting plank of the party platform: "Draw the line! Stop denazification, disfranchisement, disempowerment! Away with 2nd-class citizenship! If you want equal rights for everyone, vote FDP (formerly LDP)." Source: Haus der Geschichte, Bonn, EB-no. 1987/3/105.

> I also get so many senseless letters from Germany, assistants who had been in the Nazi party and whom I am supposed to help with their "denazification"; one of them had even reported me to the "Dozentenschaftsführer" because of my anti-Nazi mentality—but nothing happened after that. I try to help as best I can. You know, they were all just stupid boys following along, and anyway, I consider this denazification stupid and false.

One might easily think that nationalist and authoritarian figures like Gerlach or Sommerfeld had always been chauvistists and are therefore not quite representative whereas emigrés simply did not know how deeply implicated their petitioners really had been. But many other well-informed scientists were "openly annoyed at what has been going on recently [i.e., 1946/47] at some German universities under the banner of 'denazification'," as the radiochemist and codiscoverer of nuclear fission Otto Hahn put it in 1947 in articles to the *Göttinger Universitätszeitung* and the *Physikalische Blätter*. The concrete circumstance was a new wave of official

wording for his whitewash certificate on Droste: "I never had the impression that he, in his thoughts and deeds, participated in the bad side of National Socialism." (8 Jan. 1947 cited in Walker [2006] p. 124.) Just to say that Droste "criticized the crimes and mistakes of National Socialism as sharply as me" passes off Droste's prompt entry into the SA in 1933, his habit of wearing a uniform to work, and his overall "unbalanced" political attitude.

dismissals in Munich and Erlangen following American orders. The military government had the impression that these two universities had been dragging their feet, making use of the leniency granted them "to denazify their faculty members gradually, so as to avoid interruptions in the courses."[121] The chairman of the Max Planck Society in the British Zone, just recently elected on 1 Sep. 1946, and the university president at Göttingen pointed out that the "new wave of such 'official' dismissals [...] triggered very serious debates about the sense and senselessness of 'denazification' and aroused vivid memories of the whims during the 'Third Reich.'" About some of the affected persons, "it would never have entered our minds to doubt their opposition to National Socialism."[122]

The two Göttingen scholars arrogantly dismissed the competency of the denazification authorities tackling this complicated problem of evaluating political involvement. Simple allusion to their personal acquaintance with those affected sufficed for them. There was, of course, a rhetorical function for the (skewed) comparison of denazification after 1945 with the Aryanization a dozen years before. They wanted to discredit the pressure applied by the Allies for such "intellectual and moral purging." We must add that Otto Hahn was another contributor to the inflation of whitewash certificates. He too was swept up in the boycotting spirit against serious denazification attempts that created such a peculiar sense of solidarity even among formerly frosty colleagues. This experience dismayed Allied officers in charge of denazification:[123] "The proclivity of 'einwandfreie' Germans

---

[121] From *Die Neue Zeitung*, 3 Feb. 1947, 3rd ser., no. 10, p. 5: 76 "Entlassungen an der Erlanger Universität"; see also Tent [1982] pp. 92ff. and Söllner (ed.) [1986] pp. 185f. Compared with that second wave of purges at Erlangen, Kopfermann called the "purging drive" (*Reinigungsaktion*) at Göttingen "more intelligent and less wildly carried out" (letter to Charlotte Gmelin, cited in Schlüpmann [2002] p. 386). For the situation at Munich, see Walther Meißner to M. von Laue, 1 Dec. 1946 (AMPG, div. III, rep. 50, no. 1325) and von Laue's shocked letter to Meyerhof about the dismissal of 33 professors and over 60 assistants and lecturers at the Univ. of Munich, cited in Deichmann [2001] pp. 458f. He compares these dismissals to "Hitler methods" and refers to the gagging of public opinion.

[122] O. Hahn & F. H. Rein: "Einladung nach USA," *PB* **3** (1947) pp. 33–35, quote on p. 34 (this article is described further in the section: Bitterness about the "export of scientists"). M. von Laue also draws this outrageous parallel to the dismissals after 1933 in his letter to Meyerhof from late 1946, cited in Deichmann [2001] pp. 458f.; other examples of such comparisons are given in Schüring [2006] pp. 216f.

[123] According to Edward Y. Hartshorne's diary entry dated 24 July 1945 (Tent (ed.) [1998] p. 82). This sociologist was shot in 1946 while traveling on business, unfortunately leaving unfinished his envisaged book "German Universities after National Socialism. The Autopsy of an Academic Class." The contemporary concept of purging, whether or not self-imposed, is aptly typified in the title of the denazification committee for the French zone: "comité d'épuration."

to rush to the support of their colleagues who were fools enough to compromise themselves with the Nazi cause is surely one of the most startling and depressing aspects of post-Nazi German academic society."

Yet exonerating witnesses often had little incentive to clarify the role of the accused properly. Although formally "irreproachable," having never entered the party or its affiliates themselves, some were still caught up in the intricate relations of command and collaboration. After the unsatisfactory course of Rudolf Mentzel's trial, one of the summoned witnesses wrote:[124]

> The experiences in Bielefeld made me feel quite pessimistic. I have the impression that German university teachers are neither able nor willing seriously to draw a line between themselves and the people who had once led them to ruin. One is rather inclined to forgive and hide ugly actions, because there is evidently the fear that one could perhaps oneself become a little implicated in an inquiry into the circumstances.

A former coworker of Brüche in the research laboratory of AEG, who was working in a southern German tribunal after the war, was not impressed with what he observed in such judicial proceedings to determine the degree of political complicity:[125] "what I otherwise see and hear there is not particularly conducive to strengthening my respect and confidence in the German people. You have no idea how pitifully the majority of those implicated behave. But it is really quite instructive."

Because the burden of proof lay on the defendants, their trials comprised attempts at justifying themselves and denying whatever guilt they could. Concern for their personal future and the social status of those nearest to them obscured any sense of guilt and led to reflex-like defensive positions against externally imposed prosecution.[126] The statistics speak for themselves. Among the 950,126 cases tried just in the American occupation zone, ultimately only 23,776 people, i.e., less than 3 percent, were relegated into the higher categories of major offenders and

---

[124]Helmut J. Fischer to Brüche, 3 July 1949, BLM, box 3, folder 42. As described on p. 57, Mentzel was sentenced to $2\,^1/_2$ years imprisonment, which after deducting his pretrial detention led to his immediate release.

[125]Hans Mahl to E. Brüche, 15 July 1946, BLM, box 123, folder 319. According to Kielmansegg [1989] p. 33, for a while there were as many as 545 tribunals employing over 22,000 persons. A preliminary evaluation of the 11 million registration forms submitted revealed that 27.5% of the population were affected by the Liberation Law and therefore de jure 3.5 million cases awaited processing.

[126]The situation in West Germany is plausibly described by Kielmansegg [1989], esp. pp. 34f., 54f. He quotes statistics from the American zone and the later Federal Republic.

incriminated persons. But these ongoing suits stood in the crossfire of completely conflicting goals: punishment of the guilty, reconciliation with the past, and a purging of the public service for reconstruction.

Otto Hahn was initially positively disposed to the first "spontaneous denazification, which undoubtedly hit the mark."[127] By 1947, however, he had completely changed sides, fundamentally criticizing its practical implementation and denying benefits it might afford:[128]

> We [scientists] are neither politicians nor lawyers but are accustomed to regarding matters perhaps a little more calmly and rationally than other professions. We profoundly regret how the "denazification" is flipping into its obverse through the many measures, pushing true peace further and further away. We do not understand how it can take so long to distinguish finally between "criminality" and "political mistakenness"; the arbitrary muddle probably causes much of the denazification problems today including, for instance, the attacks against the science of our nation.

The blatant lack of equal treatment resulting from regional variations and the many alterations to the guidelines of denazification were criticized, often legitimately. There was the initial tendency to pass draconian punishment on collaborators without regarding many of the major offenders, who "frequently under false names and with fake passports could hide among the great number of displaced Germans."[129] Karl Bechert declined the appointment as Sommerfeld's successor at Munich and as Minister of Culture for Greater Hessen, because "the university policy in the American zone [...] is the opposite of reasonable" and he considered

---

[127] Hahn & Rein, footnote 122 above. This passage shows that in the first few months following the unconditional surrender, a positive attitude toward denazification definitely existed among the population. Disappointment about its implementation by the Allies soon changed this. Cf. also Merritt & Merritt (eds.) [1970] pp. 35ff., 79f., 304f.

[128] Hahn & Rein in *GUZ*, 2nd ser., no. 6, 21 Feb. 1947; quotes from pp. 1–2. Cf. H. Rein to O. Hahn, 4 Jan. 1947, AMPG; div. III, 14A, no. 5730, for examples of failed attempts at denazification and Rein's hard line: "Whoever has not been working as a scientist in Germany in the past 12 years has no right to pass judgment [...] What are they thinking of in England by committing such inhumanities? Are these acts of revenge? disciplinary measures? or—stupidities?"

[129] Latour & Vogelsang [1973] pp. 179f., describe denazification as "a genuine tragicomedy of errors." Particularly the actions in the American zone "have to be regarded as a moral debacle." Excerpts from the contemporary German press about a "denazification comedy," as "a protracted tragedy of justice," or a "flop and a scandal from start to finish," are given in Fritzsch [1972] p. 24. Meanwhile, in 1947, there was talk abroad that questionnaires cannot make up for missed revolutions: ibid., p. 27.

"the purging law in its present form a monstrous stupidity."[130] In May 1947, a sympathetic ministerial official gave Wolfgang Finkelnburg, who was categorized as a fellow traveler, the confidential advice to try to get employed in the British zone or preferably the French, which were both "less strict than our American one."[131] On one hand, the pragmaticism of the British was seen by critics as "arbitrariness toward the defeated"; on the other hand, the more rigorous treatment in the American zone was considered inflexible and unrealistic. The Germans were not alone, however, in questioning a justice that placed the full burden of proof on the defendant at denazification tribunals and trial courts. It was up to the defendant to find exonerating witnesses and character references to prove his or her innocence or at least argue for some mitigating circumstances. So it was guilty until proven innocent, rather than the opposite.[132] A year later this critical view had not changed. In a survey of the elapsed year 1947, Ernst Brüche seconded Hahn's polemics against the "denazification evil" with the following characterization of the prevailing mentality:[133]

> A dangerous process of disenchantment is underway that does not stop
> at physicists. Employing the bad word "renazification" would be incor-
> rect, because that would imply that initially one could have spoken of the
> "nazification" of physicists.

If, with Mark Walker, we define "nazification" as "effective, significant, and conscious collaboration with portions of National Socialist policy" (to be applied to groups, not individuals), then Brüche's impression of a profound "process of disenchantment" matches reports about a reversal of the general mood. Wilhelm Hanle (1901–1993) complained, for example, in 1948:[134]

---

[130]See his letter to A. Sommerfeld, 4 Feb. and 5 Mar. 1947, available in the Internet at www.lrz-muenchen.de/~Sommerfeld/KurzFass/02465.html and 02467.html

[131]BLM, box 1, folder 2.; on these interzonal differences see also Fritzsch [1972]. Rammer [2004] offers further examples of physicists migrating for such reasons.

[132]See Gödde [1991] pp. 64f., which also points out that the parallel classificatory and verification procedures made the confusion complete and stifled any purging impulse. Cf. also the secret assessment by the American *Office of Intelligence Research*, Report No. 4626 of 15 April 1948, published in Söllner (ed.) [1986] pp. 218f.

[133]E. Brüche: "Rückblick auf 1947," *PB* **4** (1947), issue no. 2, pp. 45–46, quote on p. 46, deadline date for manuscript submission, 10 April 1948. On 28 Jan. 1948, Brüche himself was notified that he was being charged as an activist by the tribunal in Mosbach. The very next day, the verdict was: "The person implicated is not implicated." The tribunal assessor happened to be the father of Brüche's secretary (Brüche's diary no. VI, BLM). For a definition of (re)nazification see Walker [1994] p. 81, Merritt & Merritt (eds.) [1970] p. 38, 56f., Marshall [1980] p. 673. Rammer [2004] defends a more thorough, practice-oriented conception.

[134]W. Hanle to M. Born, 5 Aug. 1948, SBPK, Born papers, 279 (I am grateful to Gerhard Rammer

It makes no sense to delude oneself. An excerpt from a Danish paper showed that it is known abroad that anti-Semitism is growing in Germany again. Many otherwise quite reasonable people here have recently become anti-Semites. The main cause is probably the numerous camps of refugees from the East of Jewish origin, people who clearly had lived under quite different conditions and cannot accustomize themselves to the local circumstances. A large part of the black market was and still is, as far as it exists, in their hands [...]. It is a hard fact that the political prestige of non-Nazis has suffered severely from the presence of these Jewish deportees. [...] and we are increasingly having to defend ourselves against anti-Semitism among our colleagues or the general public.

The revival of anti-Semitism unsettled many. It revealed that one of the core constituents of Nazi mentality had not yet been overcome.[135] Hanle's belligerent attitude is alarming, though. He evidently had no difficulty assigning his colleagues among either Nazis or non-Nazis. Unless they had an unusually extreme political record, they were able to find some former colleagues only too willing to write an exonerating reference in preparation for their hearings; and cordial congratulations followed each successful acquittal.[136] What Lutz Niethammer has termed a "fellow-traveler factory," primarily on the basis of denazification trials in Bavaria, thus also applies to our smaller sampling of scientists. So does Norbert Frei's finding of "an enormous social antipathy towards a thorough legal investigation

for drawing my attention to this letter). See also p. 89 below for another letter by Hanle; and Söllner (ed.) [1986] pp. 208f. for a similar assessment of the situation by the American *Office of Intelligence Research*, Report No. 4237 of 3 June 1947.

[135] Cf., e.g., Morris Berg to Theodore von Kármán, 26 June 1948 (cited in Hool et al. [2003] pp. 233f.) about a conference of German Jews in Düsseldorf at the time in which the complaint was raised that anti-Semitism had become worse again throughout Germany. See also excerpts from the exchange between the aerodynamicist Kurt Hohenemser and his former colleague Richard Courant, as well as with relatives and friends abroad about the prevalence and strengthening of anti-Semitism in postwar Germany, cited in Rammer [2004] pp. 557–560; Merritt & Merritt (eds.) [1970] pp. 146, 239f.

[136] See, e.g., E. Brüche to W. Finkelnburg, 4 Oct. 1946, BLM, box 123, folder 319: "My hearty congratulations on the 'fellow-traveler' distinction! Now you are rid of the main worry. You probably already know that Scherzer has meanwhile also ducked under among the fellow travelers." See also ibid., for Brüche's exonerating certificate for Finkelnburg or box 104, folder 261 for his similar defense of Otto Scherzer, dated 24 Mar. 1948.

Fig. 11: Political cartoon by M. Radler about the "denazification laundry," from the Munich satirical Magazine *Der Simpl. Kunst, Karikatur, Kritik*, Vol. 1, 1946. The whitewashing machine for black sheep bears the label "Denazificator. Patent by [Bavarian interior minister] H. Schmitt."

of the crimes of the NS period."[137]   Quantitatively speaking, it was possibly even more extreme. In search of members of the physics community who had been condemned for anything above fellow-traveler status, I come up with a mere handful of examples. The astronomer and advisor at the Reich Ministry of Education, Wilhelm Führer (1904–1974), was condemned to four years of forced labor, and the physical chemist Mentzel, who was placed in category III: lesser offenders (*Belastete*), was given a prison sentence of two and a half years. Because Mentzel's internment in Nuremberg from the end of May 1945 to 23 Jan. 1948 was deducted, he was immediately released after the court had announced its verdict.   Göttingen-trained Wilhelm Führer, classed a lesser offender by the tribunal of North Württemberg in 1949, then fellow traveler in 1950, also regained his liberty.[138] Another to come away lightly was Abraham Esau (1884–1955). He had been plenipotentiary of nuclear physics at the Reich Research Council from the end of 1943 to 1945 as well as plenipotentiary for high-frequency engineering and radar 1944/45, and had been an influential member of the university lecturers leadership.   His internment in France, Germany, and Holland 1945–48 ended with the Rendsburg tribunal classing him among the exonerated (category V, *entlastet*).[139] The physicist Eduard Gottfried Steinke (1899–1963) was among the first at the University of Freiburg to stand trial. He had joined the NSDAP in 1933 and was a member of the SA. On the strength of his party affiliations he

---

[137] See Niethammer [1972]. Cf. Mehrtens [1994] on the collaborative relations during the Nazi period and Schlüpmann [2002] p. 396 on the awe-inspiring number of 2.5 million whitewash certificates in Bavaria alone, and pp. 419–421 there on Frei and Bauer. Günter Schwarberg goes further in contending that German justice was a "murderer's washing machine": by 1986 from among the 90,921 filed suits against persons with records from the Nazi dictatorship, only 6,479 had been brought to conclusion. Roughly 84,000 had been aborted by decree without any public supervision. In the US zone more than 50% of approx. 950,000 cases ended with fellow-traveler decisions and a fine. Rauh-Kühne [1995] p. 55 discusses the paltry tribunal results by August 1949. See also ibid., p. 61 on the "out-of-hand investigative practices and abrupt changes" in British denazification. Gödde [1991] analyses the rapidly changing guidelines. According to W. Schulze in König et al. (eds.) [1997] p. 258, from among 13.2 million questionnaires filled out in the US zone, $^3/_4$ were eliminated by the Liberation Law of 1946 as "not implicated." Only 10% of the remaining 945,000 cases were indicted. Among the 1,645 major offenders, and 22,122 incriminated offenders, most were scaled down to minor offenders or even to fellow travelers.

[138] Stark is discussed in Kleinert [1983]: While the tribunal was still going on, Brüche tried to obtain information about it for the *Physikalische Blätter*, but the tribunal and the appeals court of Upper Bavaria refused his petition. On Mentzel see Rasch's article in *Neue Deutsche Biographie* (Munich); on Führer and other astronomers see Litten [1992]. The astronomer Heinrich Vogt (1890–1968) was dismissed from the Heidelberg faculty before his hearings were concluded: see Sellin [1996] p. 101.

[139] On Esau see Schmucker [1992], esp. pp. 92ff. for the postwar years, Rusinek [1998], Hoffmann & Stutz [2003], Grüttner [2004] p. 45.

had not only received his full professorship in 1937 but from 1941 also filled the high-profile office of local leader of the Nazi Lecturers League.[140] As a result of his hearing in May 1945 Steinke was initially imprisoned that summer but released again in September on instructions of the military government. In 1948, a tribunal eventually ranked him a fellow traveler and sent him into early retirement in 1950. Finding a professorship in Argentina, he stayed there until 1956. In the following year the Ministry of Economic Affairs of Baden-Württemberg accepted him as their consultant on nuclear physics. At the same time he taught temporarily at the Stuttgart Polytechnic until he managed to regain permanent employment there in 1960 as full professor of nuclear physics. He was also director of the School of Nuclear Physics at the Nuclear Reactor Society in Karlsruhe.

To recapitulate: Among a membership of 3,000 just in the *Gesellschaft für technische Physik* shortly after its founding in 1920 and about 1,200 members of the *Deutsche Physikalische Gesellschaft*, hence roughly a total of 5,000 physicists around 1945, we have a conviction rate of 1 percent.[141]

Experimentalist Rudolf Ladenburg (1882–1952) foretold this lukewarm outcome of the initially apparently so rigorous tribunals as early as mid-1947. He had left Germany in 1932 to accept an appointment at Princeton. There is an element approaching pity for the defendant in his letter to von Laue:[142]

> I still don't quite understand under what authority the so-called denazification courts operate. Over here it is said that they are German courts. But Meißner writes that the "military government," therefore the American one, decides. [...] I just read about the conviction of Joh[annes] Stark in a German emigré paper here [...] "4 years of forced labor" is too severe punishment for him, of course, but it probably won't come to that...

Erwin Schrödinger (1887–1961) also seems to have regretted the conviction of "Giovanni Fortissimo," and Max von Laue even offered his support toward reducing his sentence. Ernst Brüche agreed: "I am completely of the same

---

[140]See Fassnacht [2000] pp. 115f., and Grüttner [2004]. In a letter to Arnold Sommerfeld dated 19 March 1946 (DMM, Nl Sommerfeld, HS 1977-28/A,330) Steinke described his party affiliations in detail, requesting a character reference.

[141]The number of other convictions of physicists for Nazi crimes remains obscure. Nor do I know of any comparable figures from other disciplines. Kielmansegg [1989] p. 55 mentions that in West Germany between 1945 and 1985 the national courts initiated 90,921 investigations on Nazi crimes but only 6,479 cases ended in a conviction of the person charged.

[142]R. Ladenburg to M. von Laue, 30 July 1947, AMPG, div. III, rep. 50, no. 1158. See the letters by Annemarie and Erwin Schrödinger, 16 Aug. 1947, M. von Laue's from 16 July 1947, and Paul Gottschalk's from 10 Aug. 1947, to Arnold Sommerfeld: www.lrz-muenchen.de /~Sommerfeld/Kurzfass/01183.html, 04709.html and 02883.html

opinion as you that one ought to try to obtain a more lenient sentence. Even though Stark behaved so badly, it is not in the German interest to have a Nobel laureate in a labor camp."[143] According to Sommerfeld, one of the witnesses at Stark's trial trying to plead for leniency, Paul Gottschalk, still considered the verdict as major offender justified.[144] When an American asked James Franck to provide him with incriminating evidence for Stark's trial, he responded with a pertinent psychological analysis of the reasons why Stark had come to take such an extreme position. But neither did he want to charge his former fellow experimental physicist in a court of law.[145]

> A man of this type, who felt himself to be neglected and who believed that the modern development of science went in an entirely wrong direction, was bound to become a Nazi. The Nazis said they wanted to change everything: that is what he wanted. The Nazis believed in the bad influence of the Jews: that is what he believed after Einstein's success with the principle of relativity. He, therefore, became a Nazi at an early date, and a fervent one. [...] Nobody, even in his worst dreams, can deny that Stark was one hundred per cent a Nazi. If denazification means proving that he was not a Nazi, then it cannot be applied in his case. On the other hand, I do not think, judging from his whole attitude, that he would have actively taken part in actual crimes. His field was the destruction of the mentality of the Weimar Republic. He certainly had no influence on political matters. If, therefore, I have to understand your letter as indicating that he is in danger of being imprisoned or to be punished otherwise for his activities, I do not believe that he would deserve such a fate. He is now an old man, and it is not altogether excluded that even he may have learned a little bit from the experiences of the Nazi days. I cannot believe that he would be dangerous anymore. If, on the other hand, the question arises whether he should be made a professor again, I can only raise a warning voice.

There is a strange agreement here between German physicists at home with the otherwise so differently minded emigrés, in identifying Stark as a 100 percent Nazi (with its implications of exclusion from science and teaching). There is

[143] E. Brüche to M. von Laue, 23 Dec. 1947, BLM, box 2, folder 24. There are also copies of the verdict issued by the tribunal (9 July 1947), ranking him among major offenders, as well as Stark's petition for an appeal signed by his lawyer Dr. Bergtold, 19 July 1947, ibid., box 119, folder 308.

[144] See A. Sommerfeld to M. von Laue, 24 July 1947, AMPG, div. III, rep. 50, no. 2394, von Laue to A.V. Hill, 10 May 1951 (ibid., no. 876) and various letters regarding Stark in folder no. 1908. The engineer Paul Gottschalk, chairman of the denazification committee for the PTR in Göttingen, was one of the witnesses for the prosecution in the trial against Stark: see Kleinert [1983] pp. 19ff.

[145] Original English. J. Franck to E. U. Condon, Director of the US Dept. of Commerce, 5 Mar. 1947 in reply to his inquiry from 26 Feb. 1947; both Franck papers, RLUC, box 9, folder 1.

also the same protectiveness (merely by virtue of his having been a former colleague) against the actual imposition of legal punishment—it recalls Lise Meitner's dismissal of denazification as "stupid and false."

Ladenburg turned out to be right, incidentally. In an appeal in 1949 Stark's four-year sentence was transformed into a fine of 1,000 deutschmarks. Freddy Litten has described another of the few cases of physicist major offenders in detail. Wilhelm Müller's (1880–1968) initial verdict in May 1948 was guilty of category I. In October 1948 his charge was reduced to category II (incriminated activist, militant or profiteer) and in May 1949 he was sentenced to a year of special labor and the confiscation of 20 percent of his private wealth. An appeal in September 1949, however, found him to be an "other-worldly, prejudiced dreamer" and he was classified as a minor offender in category III and sentenced to two years probation and a fine of 1,000 deutschmarks plus legal costs.[146]

I know of no other convictions of physicists that were actually carried out. Two seriously incriminated persons (Werner Straubel in Jena and Peter Paul Koch (1879–1945) in Hamburg)[147] committed suicide right after the war. In the neighboring field of chemistry there were a few more well-publicized convictions. Major chemical trusts like IG Farben Industry (founded in 1925), or Degussa were deeply implicated in the Nazis' criminal system and the business of mass murder.[148] With its state-funded industrial development of coal carbonization and rubber synthesis, IG Farben was an important player in preparations for war implemented according to the Four-Year Plan since 1936 and institutionalized in the Reich Office of Economic Development. The director of this dye concern, Karl Krauch (1887–1968), was also Plenipotentiary of Special Issues in Chemical

---

[146]The legal grounds for the appeals verdict are reprinted in Litten [2000] p. 219: "It is extremely difficult for the panel to decide who is on the side of the law and who is not in this scientific debate. The same panel also conducted the appeals trial of the former president Stark and there already the subject of the taking of evidence was the controversy between pragmatic and dogmatic physics. One thing is certain: this controversy is conducted with personal animosity and aggressiveness and the whole scientific debate is being carried out at the political level. Objectivity on both sides suffers as a result."

[147]On Straubel, see Eberhard Buchwald to A. Sommerfeld, 25 Jan. 1946 (DMM, collection 89, 006). P. P. Koch was accused of reporting about others to the Gestapo; see Renneberg in Krause et al. (eds.) [1991] vol. 3, pp. 1103, 1110ff. For other cases of suicide, the "most contagious disease running rampant here among us," see M. von Laue to his son Theo, 25 Aug. 1946, AMPG, III, rep. 50, suppl. 7/7, sheets 43f.

[148]On this subject see, e.g., Borkin [1990] pp. 136, 207, R. Stokes in Kaufmann (ed.) [2000] pp. 398ff., Deichmann [2001] and refs. given there, esp. p. 433 on the high rate of suicide among chemists after the end of the war, as well as pp. 484ff. there on how chemists came to terms with Auschwitz and the IG Farben trial.

Production. Special *Wehrmacht* experts and Nazi bureaucrats collaborated with the dye trust in planning systematic plundering of chemical companies in occupied countries and exploitation of tens of thousands of forced laborers. One example is a giant production plant for synthetic rubber and oil in Auschwitz, in which some 25,000 laborers lost their lives. DEGESCH, a daughter company of IG Farben, produced Zyklon B, which was used in concentration camps to gas hundreds of thousands of people to death.

In August 1947 proceedings were initiated against 23 board members, directors, and department heads at IG Farben. American military tribunal IV in Nuremberg accused them of war-mongering, crimes against humanity, and membership in criminal organizations. The verdict on the IG Farben case on 29 July 1948 followed 152 days of testimony by 189 witnesses and hearing records filling 16,000 pages. Thirteen defendants were convicted to incarceration ranging from 18 months to 8 years; ten others were acquitted. These relatively mild verdicts nevertheless raised an outcry in the *Physikalische Blätter*. O. Gerhardt's lengthy article excerpted a petition by the *Gesellschaft Deutscher Chemiker* to General Lucius D. Clay (1897–1978) that follows the same strategy used in the *Persil*-certificates:[149]

> From years of working together with the condemned, we know they are honorable men. We are of the opinion that the methods used by the prosecuting authority do not satisfy the methods prescribed either prior to the Hitler regime in Germany or currently in the United States of America. We are furthermore of the opinion that the judges did not take into account the circumstances of a total war within a dictatorial state governing by methods of terror. We are unable to comprehend the severity of the sentences of imprisonment for men who are thus, in our opinion, being unfairly equated with common criminals.

By reprinting this petition along with the official statement by the German society of chemists, the DPG joined the ranks of defenders of IG Farben, who repeatedly argued that the manufacturer of poison gas DEGESCH was not a member of the dye trust but of Degussa. They carefully omitted mentioning the other inhumane services it had provided for the Nazi government, including the exploitation of prisoners in concentration camps as laborers or guinea pigs for their pharmaceutical experiments. The distinct impression remains, however, that this demonstration of unbroken self-assurance and unconscionableness was not

---

[149]O. Gerhardt: "Das Nürnberger Urteil im Chemieprozeß," *PB* **4** (1948) pp. 429–432, quote on p. 429. For background information see Deichmann [2001] esp. pp. 484ff., Schlüpmann [2002] pp. 415f., Hayes [2004], pp. 283ff.

primarily driven by concern about the fates of these "honorable men." It was more in indignation about the "discrimination of the German chemical [industry], particularly the exceptional law against the IG Farben Industry, Control Council Ordinance No. 9 from November 1945, that had been a guilty verdict without the taking of evidence and due process."[150] Otto Hahn signed whitewash certificates for two managers of IG Farben. There are interesting parallels between Hahn's and Heisenberg's opinions on the baron Ernst von Weizsäcker (1882–1951) for the so-called Wilhelmstrasse proceedings against heads of the Foreign Office. The baron's son, Carl Friedrich von Weizsäcker, was a pupil of Heisenberg.[151]

Such articles, obviously written with the aim of influencing public opinion, create the impression that the Nuremberg trials rather encouraged a sense of solidarity with the defendants. Private documents of the time convey the same message. Ernst Brüche wrote in his diary on the day of Speer's sentencing on 1 October 1946:[152] "Something inside me balks at this. I feel as never before the humiliation of Germany by this trial [...] I always found the Nazis unsavory and detestable. But these people who are now holding court certainly do not appeal to me either. Particularly today, I feel the full force of our misfortune at being entirely in the hands of the victors." After initial discussions between hard-liners like US Secretary of the Treasury Henry Morgenthau, Jr. (1891–1967), and more liberal advisors like the *German Science and Industry Committee* in London, a compromise was reached. The resulting Control Law No. 25 imposed broad restrictions that led to a consolidation of the opposition of German scientists against the Allies.[153] It came into effect on 29 April 1946, drawing a formal distinction between basic research and applied research. The former was restricted only insofar as it required apparatus that was also useful in military research and development. This formulation was rather ambiguous and its severity depended on the interpretation of the individual Control Officers enforcing it. As a result, local implementation of the law varied considerably. Applied research was affected quite seriously overall by a long list of forbidden fields of research. Such modern topics as applied

---

[150]Ibid. Heisenberg [1948] also refers to the chemical industry in addition to the optical industry in his plea for "scientific research as a necessity."

[151]See Walker (ed.) [2003] pp. 33f., Sime [2006] pp. 35f., and Deichmann [2001] pp. 484ff.

[152]E. Brüche, diary no. VI (BLM, box no. 7); cf. Merritt & Merritt (eds.) [1970] pp. 33–35, 93f.

[153]This law was drafted by the American chemist and government advisor Roger Adams (1889–1971). For the background on this law and the preliminary political decisions reached, see Stamm [1981] pp. 56, 230f., [1990] p. 888, Gimbel [1990] pp. 175ff., Beyler & Low in Walker (ed.) [2003] pp. 99f. The law was also reprinted without commentary in *PB 2* (1946) pp. 49–52. On Morgenthau's motives see, e.g., Treue [1967] pp. 15–22, Albrecht Tyrell in Hauser (ed.) [1987] pp. 57ff., and Benz (ed.) [1999] pp. 358ff.

nuclear physics, applied aerodynamics, aeronautical optimization of airfoils, etc., rocketry, jet propulsion, and other aircraft development were declared off limits. It really ought to have been no secret to anyone that research conducted in these areas between 1933 and 1945 had been strongly oriented toward the wishes and goals of the National Socialist government and had benefited much from its potential military usefulness, even though some advances came too late to be applied (such as jet propulsion developed by Hans-Joachim Pabst von Ohain (1911–1998) at Göttingen). Others (such as the uranium engine) were run on a quite low budget that had been scaled down in 1942/43. Nevertheless, not many thought it made sense to forbid these areas of research categorically. On the contrary, it was generally viewed as a strategy calculated to hamstring Germany's research potential. This suspicion was particularly persuasive when seen in light of the success these areas were enjoying in the United States, sometimes with the help of imported German specialists. Some of these recruited scientists had been procured false new identities to cover up their Nazi pasts. An example is the rocketry team headed by Werner von Braun (1912–1977).[154] Quoting Otto Hahn (and F. H. Rein) again in 1947:[155]

> Personal conversations with foreign scientists and individuals in charge of monitoring German science repeatedly fan the hope that the tiny remnant of science and research granted Germany not be completely strangulated. Experts know how easy it is to prevent it from falling into the wrong hands through control measures. But what is happening here on the other front by politicians against science sounds desperate. Wouldn't it be possible to give even a glimmer of hope of a change for the better today, two years after the end of this evil, to the many who had viewed the zones by the occupation forces of England and America as the last and only hope for an end to the Hitler regime and a return of reason to its full rights? What is the purpose of evidently trying to push these people systematically into despair and apathy? The result cannot be peace for Europe.

There are a number of letters among Brüche's papers written after the publication of this article in the *Physikalische Blätter*. One was by the spokesman for the Union of Forced Evacuee Academics of Central Germany, the solid-state physicist and

---

[154]Bower [1987] covers Operation Paperclip. Albrecht et al. [1992] discuss the recruitment of specialists in the East. They refer to 2,370 cases by name and estimate a total of 3,500 such German specialists.

[155]O. Hahn & F. H. Rein, cited in footnote 122 on p. 51 above. Max von Laue seems to have fully agreed with this article: see his letter to Lise Meitner, 25 March 1947, cited in Lemmerich (ed.) [1998] pp. 484f.

x-ray expert Adolf G. S. Smekal (1895–1959). He thought, "Hahn and Rein have grasped the situation better than the local university representatives."[156] At the time, the overwhelming majority of physicists, including the ones who had distanced themselves from the Nazi ideology (or at least proclaimed as much), regarded the situation as merely a substitution of *one* hated system of restrictions, imposed by the Nazis (e.g., to stamp out relativity or quantum theory) with *another*. Many physicists long regarded the Allies as an impediment to research, without recognizing the potential of collaborating with them, as was being done at Göttingen since January 1946 in the *German Scientific Advisory Council* of eight members. Hahn, Heisenberg, von Laue, and Adolf Windaus (1876–1959) negotiated with some success with Colonel Bertie Blount, and from October 1946 with Ronald Fraser as representatives of the British military government. The issues included licensing of scientific journals, (re)certification of scientific societies, the continuation of the PTR, and coordinating the *FIAT Reviews of German Science*. This council was not welcomed by everyone. "All very smart people, but as Laue said about the Research Council: there is much talk and consideration and then absolutely nothing happens. He is only too right."[157] The German Research Council, *Deutsche Forschungsrat* (DFR) formed in 1948 by this group of people under the leadership of Heisenberg soon merged with the refounded *Notgemeinschaft* to form the German Research Association (DFG).[158]

---

[156] A. Smekal to E. Brüche, 1 June 1947, BLM, box 119, folder 308.

[157] Ernst Brüche's diary no. VI, 8 Aug. 1947 (BLM, box 7). On the previous day he noted: "Frazer, to whom I effusively confided all my worries about the stiffness of the English, also wants to help."

[158] On these science policy and funding organizations see, e.g., Stamm [1981], [1990] pp. 892f., Eckert [1990], Hermann [1993] pp. 81f., and Carson [2002]. The constructive collaboration of Göttingen physicists with Blount and Fraser is discussed in Heinemann (ed.) [1990] part 1 and in Zeitz [2006] pp. 91, 122, 132, 140ff.; see also Rammer [2004] and the documents in AMPG, Hahn papers III, 14A, 1003.

4

RUSSIAN PHOBIA

On 13 July 1945 Ernst Brüche noted in his diary:[159]

> I woke up during the night and had to think of all the misfortune in
> Germany. About Reinhold's death, about ruined Berlin, about the terror
> we all have of the Russians, of the disinterested Americans, about Germany's
> suicide, the death of German physics, and the absolute uncertainty of our
> fate. Isn't it terrible to think: Russians in the cities in which Bach, Goethe,
> Haeckel, and whatever else their names are, lived and worked? My heart
> throbbed and tears almost welled up in my eyes.

The German propaganda machine had kindled the "terror we all have of the
Russians" when the Hitler–Stalin pact was broken. Each news report from the
eastern front was used to fan it further. The advancing Soviet troops in the former
eastern territories of Germany at the beginning of the occupation unquestionably
committed atrocious war crimes, plundering, rape, and shootings. A lethal mixture
of poor military training and pay combined with a drive to avenge the deaths of
many a friend and relative during the German assault must be one explanation
for these atrocities. Worries about the rancorous measures of troops on the
rampage were only one component of this phobia. Many Germans were troubled
about the change in the strategic balance of power in Europe. "Many also see
that the creation of a power vacuum in Europe would simply mean that in a
few decades Europe would become Russian. They evidently don't want that."[160]
Werner Heisenberg was reporting here about observations he had made during
his internment in England. He had made quite similar assessments of the situation
while the Nazis were still in power, during a few lecturing trips abroad. To the
astonishment of his hosts he attached the reasoning that Europe under German
rule was in any event better than under the Russians. With the end of the war, the
first scenario evaporated but fear of the Russians remained. The arrival of Russian

---

[159]Ernst Brüche, diary no. IV, 13 July 1945, BLM, box 7. Was he aware of the line by Heinrich
Heine: "Denk ich an Deutschland in der Nacht, bin ich um meinen Schlaf gebracht"? Cf. the entry
on 28 Sep. 1945, about the Russian "population policy," i.e., the raping of German women; cf.
Anonyma [1959] for a diary of one of these women in Berlin, and Zur Nieden [2002].

[160]W. Heisenberg to A. Sommerfeld, 5 Feb. 1946, Sommerfeld papers, DMM 1977-28/136/43.
Walker [1992] discusses Werner Heisenberg's statements during his trips abroad before 1945.

troops in their fatherland only exacerbated it.   Richard Courant summarized
the mood among the Germans he had spoken to during the summer of 1947:
"absolutely bitter, negative, accusing, discouraged, aggressive. Main point: Allies
have substituted Stalin for Hitler, worse for bad. Russia looms as the inevitable
danger." After numerous conversations in Southern Germany, Courant sketched
the following overall impression:[161]

> Fear of Russians. Bitterness against French. Rumors also of American
> mismanagement. General lack of understanding for what America actually
> does to help the Germans. Little contact between scientists in different
> towns. None with abroad, almost none with Austria ... [Criticism] of Ger-
> man administration. Small-time politicians, no understanding for cultural
> issues. University has no support from them. Complaints about zone
> competition. French do not permit some scientists to travel to other zones.
> Americans and British likewise compete for scientists and allegedly, impose
> restrictions.... [Many scientists] do not dare travel through Russian zone for
> fear of kidnapping, which sounds unbelievable but is universally accepted
> as real danger.

As we now know from various publications about the work of German specialists
in the Soviet Union after 1945, the hiring methods that the Soviets used covered the full range from persuasion to outright force. The OSSAWAKIM drive
alone recruited over 2,000 scientists and engineers with their families in October
1946. They were transported to the Soviet Union by train, where they lived and
worked in closed compounds conducting research and development in nuclear
physics, electronics and electrotechnology, optics, torpedo design, rocketry, and
aircraft research, among other projects of military relevance. The Soviet Military
Administration in Germany did its utmost to make these recruitments appear
"completely voluntary" and some physicists were pressured into publishing open
letters to that effect in newspapers and journals. At least until 1946 they appear to
have been convincing. The propaganda even alleged that there were more appli-

---

[161]The foregoing is quoted from Reid [1976] p. 261. Tent (ed.) [1998] provides other examples
from among the general population, some of them supported by testimonies about experiences
under the Russian occupation; see, e.g., pp. 15, 114, 123. Vovelle [1985] pp. 88ff. views diffuse
fear as one of the most important elements of historical sensibility. Lefebvre [1932] discusses the
collective state of fear of thieves and the enemy in 1789. The BIOS officer's profile of German
scientists in 1947, cited in footnote 49 on p. 20 above, continues: "There is also a hatred of Russia
which is almost pathological, and a large proportion of the anti-Soviet rumours circulating in the
British zone originate with the German professional classes."

cants than the Soviet Union was able to accept.[162] These so-called "volunteers" had been surprised in the early morning by soldiers surrounding their homes and demanding they and their families immediately get ready for departure. Some nevertheless managed to escape (for instance, by jumping off the train traveling eastwards). As the quote from Courant documents, the rumors soon exposed such officious statements as propaganda.

Traces of this "Russian phobia" are also found in the *Physikalische Blätter.* The second issue of volume two carried a rather conciliatory article about the "responsibility of German scientists and engineers toward Russia." A hail of protest letters descended on Brüche after its appearance containing "individual details that leave no doubt that the descriptions of the events that have appeared in the press in the British zone are legitimate. The letters mention 'kidnapping' and 'armed assault.'" But Brüche still tried to uphold balanced reporting and published a few new "positive assessments that we have received" that "consider the Russian phobia as greatly exaggerated." But the side remark "that every once in a while, one of those who are collaborating with the Russians gets devoured" did not sound encouraging.[163] The censorship of correpondence from Russia could not be overlooked and by mid-1948 Brüche's tone had become more resigned:[164]

> It doesn't help to bury one's head in the sand and refuse to see the pro-
> gressive separation of East and West Germany. The differing meanings
> Americans and Russians attach to the term democracy make further alien-
> ation inevitable right now. [...] We Germans are becoming different peoples
> on this side of the curtain as well as on that, and the differing educational
> reforms can only serve to turn the coming generations on both sides into
> strangers.

This resignation about the situation in East Germany is also reflected in the correspondence of physicists. One of Brüche's informants from the eastern zone,

---

[162]See, e.g., Steenbeck [1977] pp. 169ff., Albrecht et al. [1992], Peltzer [1995] pp. 35ff. and 102ff. on the recruitment techniques used, as well as p. 40 there for an open letter by the physicist Wilhelm Burckhardt to the *Berliner Zeitung*, 20 Oct. 1946. An officer of SMAD, who later deserted, described it as "disgustingly false."

[163]Foregoing quote from *PB* **3**, no. 1 (editorial deadline 28 Feb. 1947), p. 32: "Nochmals die Ostverpflichtungen"; cf. *PB* **4** (1948) pp. 271, 452–454: "Briefe Deutsche[r] Wissenschaftler in Rußland." Balfour & Mair [1956] pp. 40–47, 76f., describe the economic and political situation in the Russian zone.

[164]E. Brüche: "Ost und West," *PB* **4**, no. 5 (editorial deadline 15 June 1948), p. 224. On censorship, see issue no. 1, pp. 39f. Brüche's conciliatory efforts between East and West continued: see Hoffmann & Stange [1997].

a former pupil of Debye, Hermann Heinrich Franz Ebert (1896–1983), criticized the overeagerness of Germans to conform:[165]

> There is no doubt that some gentlemen—I should think over there just as much as over here—are traveling far too swiftly in the occupiers' wake. But they should not be taken for the norm. There is no sense in looking at matters through rose-tinted glasses, of course. But a different gauge has to be applied to actions taken by individuals here in the East. [...] "Don't forget us" doesn't go far enough now: When you consider what points of contact are still left, it is damned little. But as a consequence of our neutral position, we scientists are especially called on to keep in touch, even if it does involve some casualties—essentially over there again, of course.

Assessments about the situation in the western zones also changed. In March 1948 Wilhelm Hanle wrote from Giessen to the theoretical physicist Max Born (1882–1970) in Edinburgh, who had been in exile since 1933:[166]

> At the moment food worries shift politics somewhat into the background. Don't you also have the feeling that the political situation is deteriorating rapidly? We were particularly shocked about the fate of Czechoslovakia.
>
> [...] We very much welcome the imminent formation of a bloc of western nations, since we regard it as the only possibility of countering the Russian advance. But will it really be of any use? We constantly ask ourselves this question. It is widely thought that an armed conflict between West and East cannot be maintained over the long term and that then the Russians will overrun us.
>
> Many also ask themselves what they should do then. The example of some of our colleagues—like Schützen—in the eastern zone show us how little freedom there is under the Russians. I do not want to live in a totalitarian state again. I had enough from the 12 years of the Hitler regime. And then what would the future look like? Either Europe will remain Russian and western culture will decline in Europe, or the Russians will be thrown out again, just as we were first thrown out of France; then while they are retreating, the Russians will carry us off with them, of course, and we either go under then or at best end up in Russia. I found it bad enough that Hertz and other German scientists have to work in the Soviet Union, probably

[165] H. Ebert to E. Brüche, 23 Aug. 1948, BLM, box 90, folder 225.

[166] W. Hanle to M. Born, SBPK, Born papers, 279 (reference to this letter by courtesy of Gerhard Rammer). The experimental nuclear physicist Gustav Hertz (1887–1975) worked in a screened site at Sukhumi on the Black Sea on isotope separation for the Soviets before returning to Leipzig as full professor in 1954 and becoming president of the East German physical society 1955–67.

for the Russian war machine. I don't want to be on the Russian side in a battle between West and East. [...]

Many wonder, if it does come to a—perhaps only temporary—evacuation of Germany (and perhaps of Europe) before the Russians, whether one would have to retreat with the Americans. But where else can one go? And what would happen to one's family? These are the thoughts preoccupying us now.

In 1950 Hartmut Kallmann pleaded with von Laue not to go to East Berlin again to attend the 250th centennial of the "Russian Academy."[167] In his view, the Berlin Academy of Sciences was no longer an association of free scientists but "just a scientific fig leaf for the suppression by the Communists of free speech, free science, independent research, indeed independent thought in general." The presence of western scientists would only grant apparent legitimacy to the "slavery, repression and crimes," similar to the recognition of the Hitler government by other countries after 1933. It "enlarged Hitler's power and reputation and above all made any domestic opposition impossible. [...] From the Nazi era I know how terrible it is to be repressed and to see other people, who are free, come voluntarily to sing along."

Physicists and engineers who later migrated back to the West gave poignant descriptions of those years of anxious insecurity and fear, when they "learned the value of reserve and distrust as tools of survival." They were allowed to return when their expertise had become obsolete and only after it had been exploited to the fullest in ongoing Soviet projects.[168] These two supporting pillars of the mental profile of the first few postwar years: insecurity and fear of what the future might hold, existed not just in these isolated research compounds but throughout the whole of occupied Germany.

---

[167] For the following quotes see H. Kallmann to M. von Laue, 6 June 1950, AMPG, div. III, rep. 50, no. 991.

[168] An example is the reminiscences of a doctoral candidate at the Institute for Applied Mechanics at the University of Göttingen: Magnus [1993].

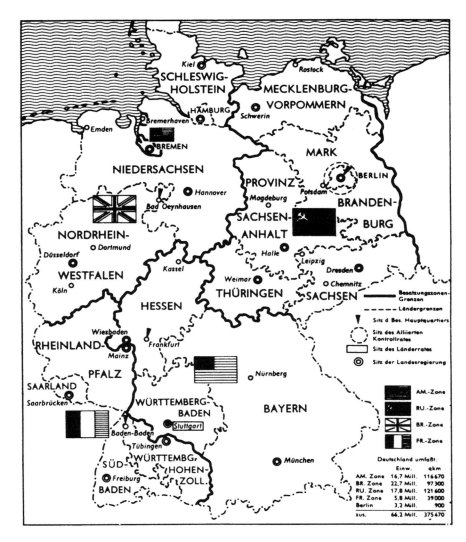

Fig. 12: Map of the occupation zones 1945–49. The heavy line demarcates the boundaries between the four occupation zones; the broken line, the borders of the provinces (*Länder*); the arrowhead marks the headquarters of the occupying forces; the broken circle, the seat of the Allied Control Council; the rectangular frame indicates the seat of the *Länder* Council; the rings, the seats of the local governments. The number of inhabitants and total areas (in kilometers) of each of the four zones are added up at the bottom right-hand corner. Source: Dollinger & Vogelsang (eds.) [1967] p. 148.

5

SENSE OF ISOLATION AND FRAGMENTATION

A painfully strong sense of isolation shines through in the last quote from Courant's report. This is another motif prevalent in texts from the period. It was not a product of the postwar era but had its beginnings in the National Socialist policy to cease all German participation in international collaborative efforts. The university president at Göttingen, Hermann Rein (1898–1953), remembered in the foreword to the just recently founded paper *Göttinger Universitätszeitung*, edited by lecturers and students: "With bitterness do we feel that intellectual isolation which had descended on German science and academe over the course of the past 12 years by the closing of the borders."[169] Hopes of reopening such ties were high, but the Allies could not fulfill them for various reasons. The Potsdam Conference held from 17 July to 2 August 1945 gave the seal of approval to resolutions reached in Yalta and announced in the Berlin Declaration of 5 June 1945 for the division of the country into four occupation zones and for the secession of German territories east of the rivers Oder and Neisse in the region of Lusatia. For simple reasons of security, movements between the occupation zones were strictly controlled, and only few people obtained—after much effort—the much coveted interzonal passports. In 1946 Otto Hahn heard from the seat of the four-powers control, Berlin:[170]

> Prof. Kallmann has been trying for a while now to obtain an interzonal passport, in order to be able to visit you, but is stymied for the time being by the bottomless bureaucracy that always manifests itself here to the fourth power. For that, one has to devote one's full working day over a period of at least a month to obtain an interzonal passport and his time does not suffice for that at the moment.

Even former victims of Nazi politics had to struggle with the strict regulations imposed by the Allied occupiers. Erich Regener, who had lost his professorship

---

[169] H. Rein: "Zum Geleit," *GUZ*, ser. 1, no. 1, 11 Dec. 1945, p. 1. The editors of this paper were initially not identified by name, so Brüche referred to the "anonymous editors" in his short positive evaluation of the new university paper in *PB*, **2** (1946), pp. 113–114. The founding main editor was the engineer Prof. Dietrich Goldschmidt.

[170] U. Martius to O. Hahn, 31 Aug. 1946, AMPG, III, rep. 14A, no. 2726 (this reference by courtesy of G. Rammer). See the account about the tribulations of travel in Germany in Scherpe (ed.) [1982].

in physics in 1937, managed to stay in hibernation until the end of the regime as director of a research station on the physics of the stratosphere in Weissenau (in the vicinity of Ravensburg, Württemberg), which was subsequently transformed into a Kaiser Wilhelm Institute. Upon receiving an invitation by the rector of the Stuttgart Polytechnic to return to his former chair, he replied:[171]

> I will come to Stuttgart as soon as possible. At the moment, the negotiations with the French authorities are keeping me very busy. But I have now made this much progress that we have the official permission to start working again, without my having lost my independence. I now have a laisser-passer for Stuttgart and also for Göttingen, where the Kaiser Wilhelm Society is now installed; only the American stamp is still missing. Provided I get some petrol, I shall come by motorbike; for the railway is still very inconvenient and time-consuming right now.

In July 1946 Ernst Brüche was stopped at the border between the American and British zones and subjected to thorough controls:[172] "The MP is diligently trying to show the Germans that they have lost the war. Right now they are in the baggage compartment. The crates are being broken open and there seems to be a prospect of confiscation. We aren't even able to be furious anymore: fatalism [...] and uncertainty about the future."

Anyone caught attempting to circumvent the border controls by crossing over to another zone without the necessary permit in the dark of night had to reckon with prying interrogations and hour-long delays extending even to days. Occasionally they fell victim to pillaging by unscrupulous border officials.[173] People sneaking illegally over the border were sometimes even shot at. In September 1947 Hermann Ebert in Thüringen called such illicit crossings into one of the western zones "life threatening." "You won't be able to believe it if I tell you that I have still not come an inch further about a passport. Distrust, harrassment ... everything imaginable. And now comes the nervousness because of the emigration of a few top people."[174] This intentional sealing off of the interzonal borders was not completely hermetic but it was perceived as yet another restriction on the

---

[171] E. Regener to R. Grammel, Rektor der TH Stuttgart, dated 30 Aug. 1945; UAS, 57/177. On 11 Sep. 1945 (ibid.) he was still waiting for his petrol ration. For Regener's *vita* see Hahn [1951], Mogun [1955], and Schopper in Becker & Quarthal (eds.) [2004] pp. 302ff. The file on his reappointment to the Stuttgart Polytechnic (UAS, SN 16/3) contains various certified statements about his political persecution during the Nazi period, "owing to his opposition to measures of the National Socialist government."

[172] Ernst Brüche, diary no. VI, 27 July 1946 (BLM, box 7).

[173] Some testimonies include Magnus [1993], p. 32, Hückel [1975] pp. 157f.

[174] H. Ebert to Brüche, 17 Sep. 1947, BLM, box 1, folder 2.

freedom of movement, this time externally imposed by the victor powers.[175] It was also seen as a muzzling of communication, by cutting off external influences in a strictly cordoned-off system. By mid-1947 Ebert was driven to the point of pleadingly reminding his colleagues in the western zones about "my anguish and my entreaty: Don't forget us, even if it causes some trouble. Indeed, precisely for that reason. We really must strive toward unification on our own initiative—even at the expense of ideal and material sacrifices. Otherwise we are ceding the advantage to enemy elements." These problems rather reactivated old confrontational images instead of breaking them down.

The zonal boundaries also greatly hampered the administration of major scientific organizations or societies at the national level. Not until 1949 were the *Notgemeinschaft der Deutschen Wissenschaft*, the committee of university presidents, or the Max Planck Society able to speak on behalf of the whole of (West) German Science.[176] The prevalence of a sense of isolation is also documented by an article that the Swiss Protestant theologian Karl Barth (1886–1968) wrote for the *Neue Zeitung* (8 Dec. 1947) issued by the American military government. Barth describes the dangers German students were exposed to in those days:

> The third danger that is menacing German students is the still almost hermetic closing off of Germany from abroad—this one form of Allied occupation policy must be emphasized in particular. I do not know what positive outcome is expected from it and what is feared by having it lifted. But it certainly contributes nothing toward the development of a new, better academic class. It is not good for a German student to live in a ghetto in which he is now secluded together with his entire nation. It is highly probable that he will have to dispense with the most important stimuli with respect to his future as long as German universities cannot be generally and freely accessible to foreigners and as long as foreign universities remain closed to German students (with a few fortunate exceptions granted after much fuss and bother). And it is also highly probable that the isolation in which German students must now conduct their studies actually eclipses a rebuilding of the autarky of the mind, giving preference to German introvertedness that has certainly not served Germany and the rest of the world well in the past.

These statements come from a person who had lost his professorship at Bonn in 1935 because of his refusal to take the official oath of office under Adolf Hitler.

---

[175] Ibid. The zonal border "signifies a heavy impediment and evokes in us here the feeling of being locked in." The following quote is from Ebert's letter to Brüche, 2 Aug. 1947. Similarly so in one of the preceding quotes also with reference to the French zone.

[176] See, e.g., Stamm [1990] pp. 889–891. On the following see Barth [1947].

(Barth had been a member of the Socialist Party of Germany since 1931 until it was banned in June 1933.) He continued to teach in his native city of Basel and had just recently revived contacts with universities and students in Germany. The other three dangers Barth listed in his article from the end of 1947 are (1) the very harsh and in many cases hopeless living conditions, (2) the failure of occupation policy, and (4) a predominance of insufficiently denazified professor-types described as "incorrigible nationalists."[177]

Allied university officers encouraged efforts to rejuvenate the teaching faculties by filling vacated positions with individuals who had previously been ousted. But there was tough resistance by university presidents, deans, and trustees ranging from timid to outright opposition. This policy was viewed as merely another form of meddling into internal university affairs already so familiar to them from Nazi policies. On the other hand, some feared it was an attempt to upset the political and social balance towards younger teachers of different mentality and norms. Geoffrey Bird suspected this attitude as the reason for the reserve he encountered from University President Smend at Göttingen and the university trustee representing the local ministry:[178]

> They were certainly not Nazis, and the only slight opposition the Rektor occasionally showed towards our policy was, I felt, motivated by the well-founded suspicion that we were aiming to break the traditional system of oligarchic government of the universities and to make possible the development of democratic ideas. He was therefore inclined to be highly suspicious of the appointment of young progressive lecturers.

Bird continued his account by recalling that the students seemed to be more receptive to his policy of democratic reform than the faculty, where internal pressures to conform were much stronger:[179]

---

[177]Barth [1947]. Murray [1949b] makes a similar observation about the "dominance at universities of reactionary professors"; cf. also Hahn's critical view, Hahn [1949] p. 3. Deichmann [2001] p. 443 cites former members of the Nazi party making up 53% of all full professors and 70% of all extraordinary and supernumerary professors holding positions at a German university latest until 1952 in the field of chemistry: "If one assumes similar relations in other subjects, we can assert that after the war German students were taught by former National Socialists."

[178]Bird [1978] p. 147. According to Bird, two university presidents in the British zone were dismissed from their offices because of their obstructionist tactics. For an analysis of the speeches made by 18 presidents at the reopening of their respective universities in 1945/46, see Wolgast [2001] pp. 287–328.

[179]Ibid., p. 150. Jacques Lacant, the French liason officer at the University of Freiburg, was less optimistic in his assessment of German students: see Fassnacht [2000] pp. 191f.

I think we had more success with the students than with the staff, as many of the latter were seldom willing to risk the disfavour of their own particular head of department and so jeopardize their academic future, unless they knew he shared their views. I am not suggesting that the majority of the professors did not want a democratic Germany, but that many were not keenly in favour of democracy in the universities.

This contrast between students and their teachers is evident in the following anecdote about a speech by the British university officer on the occasion of the ceremonial reopening of the Mining Academy at Clausthal-Zellerfeld. A simple error in translation turned an originally innocuous statement into: "Considering this faculty, you will easily understand the problems we faced in reopening this Academy." Enthusiastic applause broke out among the students in the back rows of the audience as the complexions of the professors in the front rows darkened visibly. Efforts by the university president to clear up the misunderstanding were initially futile. For the next few weeks the faculty remained cold and the students unusually cordial, because both sides assumed the university president was only trying to calm things down by retroactively reinterpreting Bird's speech.[180]

Students were not the only ones to suffer from the personal and cognitive isolation. Physicists complained about a lack of free movement inside Germany and the barriers set against outside contact.[181] The theoretical physicist Wolfgang Pauli (1900–1958) reported back in the US about "cries for help from Europe for reprints of our papers [...]. Those people over there are starved of periodicals."[182] Brüche's initiative in 1946 to have Samuel Goudsmit's (1902–1978) article on "Secrecy or science" reprinted also arose indirectly from the fear of an increasing inclination among Germans and the Allies alike to close laboratories and research institutes off from the public eye on the justification that the research was of potential military importance. Goudsmit had headed the Alsos Mission responsible for running down German nuclear physicists and collecting their files before the advancing American troops in 1945.[183] Goudsmit's call for scientific exchange among American institutions was cleverly refitted to promote the opening up of

---

[180]Bird [1978], p. 155. Rammer [2004] p. 199, points out that students and professors alike felt they were "victims of the circumstances" and rejected "separatist efforts."

[181]See C. F. von Weizsäcker's letter to A. Sommerfeld, 11 Feb. 1947; von Weizsäcker deemed "a longer interim practical" owing to "the problem of free movement" until the situation in Germany had been solved. See http://www.lrz-muenchen/~Sommerfeld/KurzFass/04893.html

[182]Wolfgang Pauli to Jauch, Lake Clear (USA), 6 Aug. 1945, in Meyenn (ed.) [1993] p. 301.

[183]See Goudsmit [1946]. Similar worries about rigid secrecy restrictions, military censorship, and the "great danger of meddling into physics by the military" are found in Pauli's correspondence from 1945: see Meyenn (ed.) [1993] pp. 285, 314, 383f. On the Alsos Mission see Goudsmit [1947].

German science to the outside world. Two years later Brüche described the mood in a letter to Goudsmit:[184]

> Thus disappointment is beginning to spread among German scientists that their hopes for an era of science without Hitler have not been fulfilled. Everyone accepts that Germany has become poor and that we must count on restrictions in every area of public and private life. But exclusion from scientific activities around the world, the tightening of research laws, the continuing dismantlings even of peaceful industrial facilities, the wranglings over Berlin, etc., are crippling any sense of initiative and creativity. We note with resignation that our colleagues abroad, who are friendly and willing to help us personally, do not see these problems or are as powerless before the political methods as we had once been before the fascist regime. Nor do we blame those among our colleagues who decide not to emigrate from the eastern zone now, but are grateful that they are staying in their positions even if it does inevitably entail compromise.

The complaints about isolation thus went beyond the issue of the seclusion of Germany or at least of the three western zones from the rest of the world. Interzonal barriers were another cause for worry. The decision to decentralize Germany and divide it up into four zones, separately governed by one of the four Allied powers along completely different guidelines,[185] had the inevitable consequence that former national organizations had to be cut up into separate entities as well. The German Physical Society was initially reopened in early 1946 as the "Deutsche Physikalische Gesellschaft in the British Zone," and the American equivalent soon followed suit. When Erich Regener began to reorganize a smaller subdivision within the French zone in Stuttgart as a regional association for Württemberg and Baden, Robert Wichard Pohl (1884–1976) protested in the name of physicists in Göttingen against "this separatist founding, which jeopardizes German unity also in the cultural sphere."[186] But if only for legal reasons, there was no other solution. When the treaty between the American

---

[184]Ernst Brüche to Samuel A. Goudsmit, 20 Sep. 1948, Goudsmit papers, Washington, D.C.; carbon copy in BLM, box 104, folder 261.

[185]The attitudes and various assessments of the situation by the four occupation forces are discussed along with the context of dividing Germany up into zones, e.g., by Hauser (ed.) [1987], Parker's contribution therein, as well as Pingel [1982] specifically on the British.

[186]See R. W. Pohl to Regener, 24 July 1946, to Schön, 12 Aug. 1946, and Ramsauer to Pohl, 13 Aug. 1946, all in folder DPGA, no. 40027, as well as Pohl to Regener, 2 Aug. 1946, AMPG, div. III, rep. 50, no. 1391. See also the minutes of the founding meeting of the DPG in the British zone on 5 Oct. 1946 and Brüche to Pohl, 6 Aug. 1946, both in DPGA, no. 40028, and Pohl to Ramsauer, 29 Jan. 1946 (ibid., 40029). The statutes of the Southwest German Physical Society contained a significant divergence: see here p. 140.

and British merging their two zones came into force in January 1947, von Laue declared in a tone characteristic of the mentality of the time:[187]

> The cultural crisis that has broken out throughout the world, and in particular throughout Germany, also interrupted at the beginning of 1945 the hundred years of work conducted by the Deutsche Physikalische Gesellschaft. We cannot continue it as a nation-wide society right now. So regional societies are temporarily taking its place.

This statement appeared in the *Physikalische Blätter*, which as of its second issue in 1947 became the publishing organ of the regional society of Württemberg and Baden. Its editor, Ernst Brüche, quite legimately saw this as "the first publicly visible link to a revived interzonal 'Deutschen Physikalischen Gesellschaft'."[188] In other instances as well, such as defining the official location of the largest review publication, *Physikalische Berichte*, the theme of maintaining Germany as a single entity reappears: "In view of the zonal divisioning of Germany, we Germans must emphasize all the more energetically that Berlin remains for us the capital city of the Reich. Relocating the Physikalische Berichte to the southwest of our fatherland would, under the prevailing circumstances, appear as just a piece of that retreat from the eastern territories that we unfortunately have to observe in so many other developments."[189]

In the British zone, Max von Laue and Walter Weizel (1901–1982) chaired jointly over a membership of 485 of the regional association of Lower Saxony, Rudolf Mannkopff (1894–1978) was managing director, and committee members included Erich Bagge (1912–1996), Hans Kopfermann, and Fritz Sauter (1906–1983). Hans Gerdien (1877–1951), Otto Hahn, Lise Meitner, Gustav Mie

---

[187] Max von Laue, *PB* **3**, Jan. (1947) p. 1. In a retrospective on the centennial celebration of the DPG on 18 Jan. 1945, Ernst Brüche refers to the "modest celebration in the sixth year of the struggle between nations [*des Völkerringens*]," playing down the world war by his choice of words: both citations in Klaus Schlüpmann [2002] p. 448. This "rhetoric of minimization" is similarly employed in the society's correspondence after 1945: Schüring in vom Bruch & Kaderas (eds.) [2002] p. 454, Sachse [2002] and Vogt [2002] pp. 97ff., 124, and 129ff. on reparations.

[188] Cited from the notice "Der dritte Jahrgang" in issue no. 1, *PB* **3** (1947) p. 32. There we also find an answer to the inquiry about the delayed founding of a physical society in Hessen. According to a letter by Kurt Madelung from 22 Feb. 1947, an application for a permit for the regional association had been submitted but nothing more had been heard since from the responsible authority. There were reservations about having the *PB* be the official publication for all physical societies in Germany. See E. Regener to von Laue, 20 Jan. 1947, AMPG, div. III, rep. 50, no. 2391: "there is not much support for it anymore, because we do not consider the level of the Phys. Blätter particularly high, especially concerning the foreign news reports, that make up the majority of the communications."

[189] M. von Laue as chairman of the DPG in the British zone to C. Ramsauer, 4 Nov. 1946.

(1868–1957), Max Planck, Ludwig Prandtl, Hermann von Siemens (1885–1986), Arnold Sommerfeld, and Jonathan Zenneck (1871–1959) were honorary members. In Württemberg-Baden, Walther Bothe (1891–1957) presided over 160 members; in Hessen, Georg Madelung (1889–1972) over 82; in Bavaria, Gerhard Hettner (1892–1968) over 162; in Rhineland-Palatinate, W. Ewald in Mainz over a membership of over 46; and in Berlin, Carl Ramsauer over some 100 members. The first commonly held meeting of all the regional societies took place from 11 to 15 October 1950 in Bad Nauheim, where it was resolved to consolidate into the *Verband deutscher physikalischer Gesellschaften*. In 1963 the national association readopted its former name, *Deutsche Physikalische Gesellschaft*.[190]

The *Deutsche Geophysikalische Gesellschaft* underwent a similar superficial regionalization after its dissolution in 1945. On 20 November 1947 a local geophysical society was founded in Hamburg "because the necessary permits for it are easier to obtain from the responsible military government,"[191] two years later to be reunited again with its old name. More was involved than a superficial reshuffling of names. It was a conscious hide-and-seek with the Allies, an expression of another element of the mentality of this period: A disingenuous urge to hide the true state of affairs from unwelcome monitors as soon as touchy points like topics of research or one's own past were involved. A reminiscence by a coworker of Peter Adolf Thiessen (1899–1990) is relevant here. Under Thiessen's directorship until 1945, the Kaiser Wilhelm Institute of Physical Chemistry had been chiseled into a model Nazi organization. After the war Thiessen worked for ten years in Sukhumi in a research program on methods of isotope separation, and from 1956 he directed the Institute of Physical Chemistry of the Academy of Science in East Berlin. During the GDR period, Thiessen confessed to his former deputy:

---

[190] The inauguration of the DPG in the British zone is covered in *PB* **2** (1946) p. 16. The first convention of physicists took place in Göttingen 1946: ibid., pp. 178f., see also Rammer in Hoffmann & Walker (eds.) [2005]. On the rules of procedure see *PB* **3** (1947) p. 29; on Bavaria and the Rhineland-Palatinate, *PB* **4** (1948) p. 124. Shortly after the founding of the Federal Republic of Germany, von Laue was no longer willing to apply to the military government for permission on matters concerning the DPG: See von Laue to Bothe, 13 June 1949 (AMPG, div. III, rep. 50, no. 330): "We Germans no longer need to ask the occupying forces about such matters of purely German concern, especially considering that the union of scientific societies certainly does not fall under the occupation statute." The general history of the organization after 1945 and the first postwar conferences are treated by Walcher [1995] as well as by Rammer in Hoffmann & Walker [2005]. Brüche's special concern for East German interests is discussed in Hoffmann & Stange [1997].

[191] See M. Koenig in H. Birret, K. Helbig, W. Kertz & U. Schmucker (eds.) *Zur Geschichte der Geophysik*, Berlin & New York: Springer, 1974, p. 5. The procedure was similar for other societies, see Stamm [1990] p. 888.

"My dear Linde, we played foul with the Nazis and we are playing foul with the Communists in exactly the same way." I would not agree with Hartmut Linde's emendation, "To him only science was important."[192] But a discussion of personal attitudes and convictions would lead us too far astray.

Fig. 13: Two officers of the Research Branch of the Military Government in front of the Auditorium of the University of Göttingen, at that time the headquarters of the British Military Government in Göttingen; c. 1947, Source: Stadtarchiv Göttingen.

---

[192] Quoted from Guntolf Herzberg & Klaus Meier: *Karrieremuster*, Berlin: Aufbau Taschenbuch Verlag, 1992, p. 21. I owe this reference to Jens Jessen. For more on Thiessen see Eibl [1999].

6

BITTERNESS ABOUT THE "EXPORT OF SCIENTISTS"

Besides the lack of new faces and ideas in barricaded Germany, there was the brain-drain problem: the departure of significant numbers of highly qualified researchers and teachers to other countries, whether voluntarily as emigrés or through forced recruitments. Specialists in the fields of weaponry or military technology such as aerodynamics, rocketry, or nuclear research were particularly sought after. In December 1946 the electron physicist Gottfried Moellenstedt (born 1912) noted in dismay: "otherwise, it seems as if another landslide is occurring in German science, in that appointments to America are descending on leading full professors of physics."[193] Things soon got worse. At the end of February 1947 the recent Nobel laureate Otto Hahn drafted an appeal together with the president of Göttingen University, Hermann Rein, under the heading "Exportation of scientists to America" and published it in the *Göttinger Universitäts-Zeitung*. Brüche adopted it for his *Physikalische Blätter* under the less polemical title "Invitation to the USA."[194] As an introduction, a few American newspaper reports about the acquisition of German scientists and engineers are reviewed. The United States purportedly made a saving of about 1 billion dollars as a consequence.[195] They then lament: "Science and scientists—without even mentioning German patents—are being described and treated as objects of 'reparation.' There is no question that things are going on that are probably unique in the history of science

---

[193]G. Moellenstedt to E. Brüche, 16 Dec. 1946, BLM, box 123, folder 319. See here also the quote on p. 20 above.

[194]Hahn & Rein [1947]. Cf. Hahn's considerably milder draft in AMPG, div. III, 14A, no. 6194 and a list naming 17 physicists and chemists with invitations to the USA for at least 6 months, ibid., no. 5730, dated 23 Jan. 1947. A letter by James Franck to Hahn from 4 Oct. 1947 (Franck Papers, RLUC, box 3, folder 10) reveals that Hahn evidently tried unsuccessfully to have this article published in the *Washington Post*. Rein's role in the context of military research on flight physiology is discussed by Beushausen et al. in Becker et al. (eds.) [1998] pp. 236–241, 252–257 and Sime [2006] pp. 33f.

[195]Cf. *Neue PB* **2** (1946), issue no. 2, p. 20 on 130 German scientists and engineers participating in military research in the USA. In issue no. 8, p. 260, E. Brüche notes in the editorial column a saving of 2 million dollars per invited specialist. A nine-page typed list from 1947 of German scientists and engineers with "USA obligations" seems not to have been published. It is signed by Br[üche], Hahn Papers, AMPG, div. III, 14A, no. 6198.

and are beginning to stimulate considerable bitterness among that relatively small group of those responsible for representing science in Germany."

Bitterness is a word that recurs frequently in the sources and constitutes one of the components of the mentality profile we are assembling.[196] After complaining about a new wave of dismissals, cited above, Hahn and Rein continue:[197]

> The official invitations to the USA mentioned at the beginning arrive in this new, unsettled, bitter situation, triggered by the threat to the work of renowned researchers and to the livelihoods of their families. As much as we wish that these colleagues in particular, who in our opinion have been unfairly affected, may continue to work in a better atmosphere more favorable to science, we very much regret that they are condemned here and dismissed from their positions—while over there, they are sought after because of their expertise. They are unfairly blamed here by the public as deserters, over there they are viewed by leading scientists as unwelcome intruders.

> Most of the older professors leave Germany very unwillingly, because they feel that their place is here. Necessity compels them, because their livelihoods and working opportunities in their own country are taken away from them or else they are left in a constant state of fear of such an occurrence. All this, after our having experienced well enough what it means to replace competence with "politically irreproachable" dilettantes. But more depresses these men: the awareness that it is evidently not a

[196] For instance in Hahn [1949] p. 3: "It is certainly understandable that the factory dismantlings still taking place four years after the capitulation are being greeted with bitterness, particularly among the academic youth." This bitterness outlasted the time frame of the present analysis. The following example taken from a letter from 1957 by Adolf Baeumker (1891–1976) to Walther Gerlach incorporates a strong sense of duty characteristic of civil servants. Baeumker had been serving as coordinator of the research headquarters in the Reich Ministry of Aviation since 1942 (cited in Schlüpmann [2002] p. 447): "I shall depart from this life with deep bitterness about the thoughtlessness of the lawmakers (who did not incorporate any hardship clause) and about the administration of the new state (who allowed the highest Federal Court in Karlsruhe to reject the eligibility of former civil servants of the Reich to benefits of the civil service). My heartfelt love for my dear, old fatherland will remain unchanged as before. I will pray for it as I always do." Balfour & Mair [1956] p. 152 mention "a great deal of bitterness" with reference to the bad food situation during the winter of 1946/47.

[197] Hahn & Rein [1947], quote on p. 34. Reparations are discussed in Balfour & Mair [1956] pp. 80ff., 131ff., 144ff., 162–168. A similarly adventurous historical parallel is drawn between the stage shortly after 1933 and the stage following 1945 by F. H. Rein [1945] p. 7: "Just like in 1933, we hear the question: 'Where were you?—You were afraid and remained quiet [...].' One often hears and reads exactly the same words as in 1933. There is as little lack of suspicion and denunciations as then. The same sorry business is starting all over again."

matter of an honorable appointment to an independent American research institution or university of some rank but (at least according to the American press) forms a part of the "reparations." Centuries ago, princes sent their countrymen away as plantation workers or soldiers. Today, scientists are exported.

After seeing repeated parallels drawn between Allied science policy and measures taken by the Nazi regime, this time we have a potent comparison with absolutist governmental practices. Ernst Brüche was fully aware that by reprinting the appeal in the *Physikalische Blätter* he was risking the loss of his publishing license from the military authorities. That is why he added a footnote to the article pointing out that he was just following the same procedure as had recently been done in the *Neue Zeitung*, a paper issued by the American military government. It had reprinted text from *Foreign Affairs* with the comment: "We believe we are contributing better toward orienting public opinion by granting generous hospitality to a variety of ideas than if we identify ourselves with a particular direction. We do not assume responsibility for the views published here, but we do assume responsibility for having given them an opportunity to be published." (Ibid.) Formally, Brüche was in the clear, even though he was of one mind with the two authors. We gather this from his annual retrospective in the *Physikalische Blätter*. From the safe vantagepoint of a year's interval, he wrote that Hahn and Rein "had found clear and appropriate words about the professional obligations of German scientists in the United States, the essential points of which any physicist would agree with."[198]

Not everyone was pleased about this article, as Brüche had anticipated, especially scientists who happened to be away on such visits to the US or England right then. They did not like to see this personal distinction suddenly reduced to a form of reparations payment. Georg Joos (1894–1959) complained to Sommerfeld about the "presumptuous tone" of the *Göttinger Universitätszeitung*, foretelling that it would be the source of the "next debacle."[199] James Franck also carefully

---

[198] E. Brüche: "Rückblick auf 1947," *PB* **3** (1948) p. 45 (published on 10 May). He concurred likewise with the contribution by W. Kliefoth, on "Physicists as reparations," *PB* **2** (1946) pp. 369–370. Kliefoth had asked: "how should the 'dismantled physicists' be reckoned [against the reparations debt]?" Cf., e.g., Arnold Eucken's letter to A. Sommerfeld, 28 Aug. 1947. Eucken regarded Clusius's emigration as a "catastrophe" for German science: /~Sommerfeld/KurzFass/04673.html Likewise Heisenberg [1948] who feared: "the big danger is that the very best emigrate."

[199] G. Joos to A. Sommerfeld, 30 Apr. 1948, in DMM, collection 89, 009. I owe this reference to Oliver Lemuth. His study, coauthored with Rüdiger Stutz, in Hoßfeld et al. (eds.) [2003] on physicists and chemists in Jena, goes into greater detail about Joos. Cf. the letter by another physicist with an assignment in the US quoted in *PB* **3** (1947) pp. 62f. It offers a justification for the "although not explicit, nevertheless clearly perceptible objection to entering into an employment contract" of

expressed reservations about it in a letter to his former close friend Hahn:[200]

> By the way, I am not quite sure that I completely agree with you. I don't
> love the "expert-transfer" method for various reasons, that's true. On the
> other hand, under the present conditions that unfortunately happen to be,
> I am not sure whether at the moment Germany doesn't perhaps have more
> academics than it can feed; if this is the case, there is a good side to this
> export, noxious though it is to me in many respects.

Hermann Francis Mark (1895–1992) also wrote Brüche. He had formerly taught
at the Technical University in Karlsruhe followed by a professorship in chemistry
at Vienna before his emigration in 1938 and subsequent appointment in 1940 at
the *Polytechnic Institute of Brooklyn*. Mark wished Brüche had seen the amusement
on the faces of his colleagues (including Brill, Doneth, and Kallmann) when he
showed them the article about their purported exploitation:[201]

> After a few months, most of these colleagues accepted teaching assign-
> ments at various universities, some others stayed with the Army or Navy
> and are working in research laboratories there, which in general are quite
> well equipped and are being reasonably managed. Others preferred to
> return to Germany. If these latter are now spreading all kinds of inaccu-
> rate news about the fates of former German scientists in America, it is
> probably partly attributable to injured pride, partly also to that unpleasant
> nationalistic arrogance which has already wreaked such havoc in the past
> and unfortunately still is hurting German interests. I personally would
> recommend all my German colleagues to have a look at the conditions here
> once with an open mind.

Rudolf Ladenburg's letter to Max von Laue, written right after he had read the
article, adds another aspect:[202]

> Otto Hahn's article in the Göttinger Universitätszeitung draws a link be-
> tween dismissed German professors and American offers of professorships
> in the USA. Do you know of anyone from among our colleagues who was
> first dismissed & is now moved away to the USA? I only know that West-
> phal is in the USA & that Joos left for here at the beginning of June [...].
> Westphal was, as far as I know, never dismissed & Joos was re-engaged

the kind. Two years later there were still reservations about it: see the exchange of letters between
the USA and Germany in *PB* **6** (1950) pp. 190ff.

[200] J. Franck to O. Hahn, 4 Oct. 1947, Franck papers, RLUC, box 3, folder 10.

[201] H. Mark to E. Brüche, 23 Nov. 1949, BLM, box 2, folder 24.

[202] Ladenburg to von Laue, 30 July 1947, AMPG, div. II, rep. 50, no. 1158, cited in Schlüpmann
[2002] pp. 426f. Meitner's letter, excerpted on p. 25 above, provides more information on Westphal.

after a short period of time. In his case, the dismissal was based on an (unforgivable) confusion of names.

Ladenburg sensed an ulterior motive in the article in correlating the dismissals with the importation of German scholars for exploitation in American weapons development. By insinuation the article leaves an unpleasant aftertaste of duplicitous ethics and unscrupulous self-interest.[203]

Wolfgang Finkelnburg (1905–1967) is one example. He was vice-chair of the DPG until 1945 and appointed regular extraordinary professor of physics at the Reichsuniversität in Strasbourg in 1942. After the war, his year as acting head of the University Lecturers League in Darmstadt 1940/41 caused him trouble. His party membership (dating to 1937) was cited as the reason for his dismissal as vice-chair of the DPG.[204] He was categorized as a fellow traveler. In May 1947 a university expert of the Minister of the Bavarian Province informed Finkelnburg that he "might by law be fully reemployable, but that practically, reengagement in a position on the university budget was as good as impossible. It may look better a year from now. For the time being, at least, we must limit ourselves to rehiring 'unimplicated' and 'exonerated' persons; otherwise we would throw away their trust, as we have been informed."[205] Fear and preemptive obedience continued to reign at the ministerial level as well, even after the demise of National Socialism. It was just marching to a different tune. When a scathing article appeared in the *Schwäbische Landeszeitung* signed by G. von Studnitz in issue no. 41 (1947), under

---

[203]Rein's own ethics leave something to be desired, as Hohenemser's case reveals. Hohenemser had been dismissed from the University of Göttingen in 1933 for anti-Semitic reasons but was able to find employment in industry until the end of the war. When he attempted to return to the university after the war, his permission to teach was retracted in 1946 during Rein's presidency, leaving him no alternative to emigration. This case exemplifies the university president's double standards: He was willing to solicit highly qualified former staff members, but only if their personalities were amenable to him. See Hentschel & Rammer [2000], [2001]. On the other hand, the names of a few chemists and members of the board of IG Farben, who were convicted as war criminals (Bütefisch, Krauch, and von Schnitzler), appear on a list compiled by the secret service for rehiring in the USA following their dismissals; see Deichmann [2001] p. 447. Rubinstein [1947] p. 18 points out another aspect of this transfer of know-how: "the importation of Nazi scientists is accompanied by the importation of Nazi ideas."

[204]See the notice by Ramsauer "On the current composition of the boards of the Deutsche Physikalische Gesellschaft and of the Deutsche Gesellschaft für technische Physik, Berlin" 18 Dec. 1945, DPGA. It reveals that Ramsauer initiated this measure himself, "in order to make the board effective again" and acceptable to Allied Control officers. This measure was quite typical of the period: cf., e.g., A. Schildt in König et al. (eds.) [1997] pp. 225f. on the replacement of persons holding high office at almost all universities, often just days after the capitulation.

[205]Hans Rheinfelder to W. Finkelnburg, 27 May 1947 copy in BLM, box 1, folder 2.

the heading "Scientists as war booty," castigating anyone deciding to leave for the US as traitors, Finkelburg sent a letter in protest to the editor under the title "Do German scientists have to emigrate?" Finkelburg portrayed the "fellow-traveler problem" from the point of view of someone affected, who saw no alternative to leaving for the US. He generally agreed with the article by Hahn and Rein but offered another reason for why people felt obliged to accept special job offers from the USA, even though they would have much preferred to stay in Germany:[206]

> We youngsters have lost confidence in regular university appointments. [...] Calls like the one in Mainz inevitably give us the impression that merit as a researcher and university teacher has nothing to do with it, just a clean political record taken to the extreme and those notorious "connections." [...] It seems to me that people are far too timid again about just wanting to conduct science. That is why we have to pack up for good and go abroad!

The DPG refused to assume the role of advocate of persons dismissed for political reasons. Max von Laue heard about Finkelnburg's letter from Otto Hahn and answered in May 1947: "This colleague of ours has always been a very good prosecutor of his own affairs. He is in this case, too. But I can only reply to him: If a university faculty turned to the D.Phys.Ges. about appointment matters, he would surely get a response. But none have done so thus far."[207] So it was not the DPG as an association that decided but the network of influential referees and gray eminences behind the scenes, as Wolfgang Finkelnburg soon realized: "in Germany one only has prospects when one has been ordained at Göttingen."[208]

But the correspondence by physicists employed in the US or Great Britain after 1945 reveals that a lack of opportunities in Germany was not the only

---

[206]W. Finkelnburg to O. Hahn, 18 May 1947, AMPG, div. III, 14A, no. 926. Likewise his letters to E. Brüche from 8 March and 5 June 1947, BLM, box 1, folder 2, where he almost laments having to sell himself "to our competitors in industry by accepting offers in the USA." Finkelnburg returned to Germany in 1952 and worked in the Schuckart facilities of the Siemens electric company in Erlangen. In 1955 he was awarded an honorary professorship there. See Beyerchen [1980] pp. 238ff., Hentschel (ed.) & Hentschel (ed. asst./trans.) [1996] p. XXIV, Grüttner [2004] pp. 47f., and further references there.

[207]M. von Laue to O. Hahn, 24 July 1947, AMPG, 14A, no. 925. See also the exonerating certificates by Ramsauer and Brüche for Finkelnburg's tribunal, dated 12 Dec. 1945 and 29 May 1946; both are copies among correspondence between Brüche and Finkelnburg in BLM, box 123, folder 319.

[208]W. Finkelnburg to A. Sommerfeld, 8 Aug. 1947, Sommerfeld papers, DMM, online access: http://www.lrz-muenchen/~Sommerfeld/KurzFass/04691.html Robert Wichard Pohl was very influential regarding such appointments; on the importance of "collegial networks" see Rammer [2004], Schüring [2006] pp. 134f., 158, 202.

element at work. Better economic and working conditions in science were a primary motivation in their decision to take such a step. The geophysicist Max Diem (born 1913) at Karlsruhe commented on Brüche's "if not explicit, certainly clearly perceptible objection to accepting employment contracts in USA" in early 1947 by pointing out the possibility of "working as a scientist under conditions that we in Germany will not be able to have for the coming years, perhaps even for decades. On the other hand, the living conditions are so substantially better that even just a brief stay means an important relief for us."[209] Shortly before his departure to the US in May 1947, Georg Joos added another reason: "[I] am looking forward to not having to see some colleagues anymore."[210] A year later he justified his decision not to return to Germany for the time being with the following argument: "The population count has to be reduced or else we shall all just waste away like the Indians."[211] Wilhelm Hanle, who decided against going abroad, summarized the arguments pro and contra emigration to the US in 1947 as follows:[212]

> One could be of the opinion that under the circumstances, especially right now, one should not leave Germany in the lurch, particularly if one has a clean record. But one could also be of the opinion that 20,000,000 have to emigrate anyway; that one is doing a good thing by making room. One could also be of the opinion that this state, by continuing to commit the same old mistakes and to a large part by continuing its false university policies, does not deserve our continuing to serve it. [...] It also certainly looks good when one of our major authorities proclaims that he will not leave Germany, his homeland, in its time of need, even when a good offer is made to him to leave for the USA.

[209] M. Diem to E. Brüche, 11 Feb. 1947, BLM, box 1, folder 2. Brüche himself categorically rejected any offers to go to the US, even when they promised a team of coworkers. See, e.g., Brüche to the director of AEG, Dr. Boden, 4 Feb. 1947 (ibid.).

[210] G. Joos to Malsch, 8 May 1947, copy in BLM, box 119, folder 308. He was thinking particularly of Lenard, Stark, and Malsch. Finkelnburg had written an article for the *PB* on the Munich Synod, a debate arranged to settle a dispute over the theory of relativity (cf. Hentschel (ed.) & Hentschel (ed. asst./trans.) [1996] doc. 110). Malsch had been portrayed as spokesman of the Aryan physics faction. He protested against this designation, calling himself "neutral." Joos and Finkelnburg contested this, whereas von Weizsäcker indirectly confirmed it. See ibid., folder 308 on the correspondence between Brüche and Malsch in March 1947 as well as von Weizsäcker on 3 and 10 April 1947 and with Finkelnburg on 12 May and 5 and 7 June 1947, box 1, folder 2.

[211] G. Joos to A. Sommerfeld, 30 April 1948; DMM, Sommerfeld papers or www.lrz-muenchen.de/~Sommerfeld/KurzFass/03074.html

[212] W. Hanle to E. Brüche, 18 March 1947, BLM, box 119, folder 308. After describing the predicament of a person who had endured 10 years' imprisonment, this quote continues with the advice to emigrate: see here p. 150.

In an effort to slow the trend of emigration, von Laue even resorted to exaggerated polemics. In an interview with a reporter of the *Darmstädter Echo* in mid-1949, soon after returning from a trip to the US at the end of July, von Laue explained that he and his wife "were served particularly hospitably in the stores" when their identity as Germans was revealed by their questions and requests.[213] However,

> many researchers employed in military research laboratories silently yearn for the peace of their former quest for truth, even though collaboration on the major projects does attract American scientists, of course, who are generally not very well situated financially. But some are compelled to find compensation in civilian industrial research for lack of state funding. [...]. Some of those who had sold themselves to the USA in 1945 are doing quite poorly. They don't live much better than prisoners, receive 6 dollars pay—per day—and may only leave their assigned location with special permission.

As had been the case with von Laue's earlier statements, this publication too triggered harsh criticism abroad, for instance from a former close friend of his. The crystallographer Peter Paul Ewald (1888–1985) had resigned his office as university president of the polytechnic in Stuttgart in protest against the notorious racist Law for the Restoration of the Professional Civil Service in 1933, and had emigrated four years later. From Brooklyn, New York, Ewald wrote his former colleague that the interview had caused "astonishment and repudiation" and drew a completely false picture of the situation. The subsequent exchange of letters could not change the fact that von Laue's friends in the US were "very disappointed and disconcerted, particularly with respect to the effect it could have in official places, who are conducting the exchange of students and lecturers mentioned in the last sentence." Max von Laue's polemics had done some damage to the touchy business of normalizing international scientific exchange. The contributions by Clemens Münster (1906–1988) show that it could have been otherwise. Münster was a physicist employed in private industry who became editor of the liberal *Frankfurter Hefte* after 1945. From 1949 to 1971 he was chief editor of culture and education and from 1954 on, also television director of the channel Bayerische Rundfunk in Munich.[214]

---

[213]M. von Laue [1949c] with subheadings by the editor: "No trace of hate" or "Ready to serve." For comparison, cf. von Laue's letter to the *Darmstädter Echo*, 5 Jan. 1950, AMPG, div. III, rep. 50, no. 2339. For the following: Ewald to von Laue, 23 Nov. 1949 and 19 Jan. 1950, ibid., no. 562.

[214]His description of the situation, when compared with von Laue's dramatization, was much more relaxed: see Brücher & Münster [1949] esp. pp. 341f.: "There has been much talk about the 'dismantling of German scientists' since the end of the war." See the biographical entry on Münster in the *Deutsche Biographische Enzyklopädie*, suppl. vol. 11 (2000) p. 133. Hildegard (Hamm-)Brücher later became a well-known representative of the FDP in Federal Parliament.

The exodus persisted nevertheless. Between 1950 and 1967, the DFG counted a total of 1,400 migrant scientists lost to Germany.[215] Bitterness also characterizes the tone of various reports regarding "The situation in the area of patenting in Germany."[216] Just like after World War I, all German patent rights were confiscated and released as property of the Allies "for mutual use" (according to an announcement in the *Tagesspiegel* of 10 Aug. 1946, they numbered over 100,000). In April 1946 the US Department of Commerce published an appeal urging American companies to use the new possibilities to full advantage: "Never before has American industry had such an opportunity to acquire information based on painstaking research so quickly and at such low cost. This is part of our reparations from Germany in which any American may share directly."[217]

A decision on domestic patents and new patent applications was only reached later on, which led to a considerable backlog of pending claims. In August 1947 a law finally came into force in the US making it possible (under certain conditions) for German citizens to file patent applications in Washington again. This had been feasible in France since April 1946, but it took two years longer (until April 1948) for it also to be possible in England. The protracted legal uncertainty and the impression of having been deprived of legitimate rights strengthened the widespread sense of victimization, even when this "plundering" took place with explicit reference to identical practices Germany had itself followed in the occupied territories until 1945.[218] The overall extent of the reparations was mostly overrated at the time, however. Later investigations by the United Nations and other independent experts found that reparations payments from the western zones totaled only about 5 percent of the available capacity, or, expressed in monetary units, material goods valued at under one billion marks. Altogether,

---

[215]See Stamm [1981] p. 274 and Schüring [2006] p. 322.

[216]Thus the title of a longer article by the engineer and expert on patent law H. G. Heine: "Die Lage auf dem Patentgebiet in Deutschland," *PB* **3** (1947) pp. 387–389. Cf. also "Das Patentwesen nach dem Zusammenbruch," *PB* **2** (1946) pp. 169–174, Fr. Frowein: "Schutzrechte im Ausland," *PB* **5** (1959) pp.15–17 and Gimbel [1990] on the "Exploitation and plunder" by the Americans, esp., pp. 63, 173 on patents as well as Farquharson [1997] on the British and on the right of perusal granted even to German companies, who thus gained easy access to the developments of their direct competitors. For a critique of Gimbel's thesis see Volker Berghahn's contribution in Judt & Ciesla (eds.) [1996].

[217]*Washington Post*, April 1946, quoted from Hermann [1993] p. 79. Interesting examples are given in Hans Stieber: "Zehntausend Geheimnisse," typescript in BLM, box 104, folder 261, based on the English original article: "Secrets by the thousands" in *Harpers Magazine*, Oct. 1946.

[218]For statistics and commentary see Treue [1967] pp. 89f., Gimbel [1990] and Judt & Ciesla (eds.) [1996], in the latter, particularly Jörg Fisch, pp. 11ff. on legal aspects and Matthias Judt, pp. 28ff. on economic aspects.

Germany spent approximately $16.8 billion on reparations and costs of occupation between 1945 and 1953 but got credits in the amount of $1.5 billion in the same period. During the following decades (until 1988) it spent another $15.4 billion for public and individual reparations, occupational costs, and services. The eastern zone satisfied the Soviet Union's demand for $10 billion in reparations to about 94.1 percent. This explains the more successful economic recovery of the western zones which reached their 1936 levels of production already by the end of 1949.[219]

Reparations were a problem affecting the society as a whole but the demoralizing effect had specific repercussions on universities. In a letter from March 1948 Wilhelm Hanle reported to the emigré Max Born about a change of mood at the University of Giessen. Catching the small but determined faction of non-Nazis off guard, a former "March casualty" (as members of the National Socialist Party were called who had joined only after it had proved victorious at the polls in March 1933) carried through his candidacy as vice-president of the university. How could he get that far?[220]

> Why did we [i.e., the group of 6 non-Nazis] lose prestige? Our intervention for the Jews under the above-described circumstances [of a black market dominated by formerly persecuted Jews, cf. here the quote on p. 55] is naturally only part of the reason. Much was developing in a downwards trend in the last few years. The state had no control over the demoralization. The black market was flourishing. The state watched while good, hard-working citizens suffered and starved. All 4 occupying forces acted—to put it mildly—very clumsily. Perhaps, after ourselves having acted like that for 13 years, we have no right to criticize others. But it would have been wiser not to leave Germany on tenterhooks like that. How much political prestige did we non-Nazis lose, just from the military government's dismantling of a harmless soap factory in Giessen! The Nazis certainly are guilty of wartime measures causing a soap factory somewhere in France to be destroyed, and France is now demanding compensation. And yet it is not sensible to continue to [dis]mantle strictly civilian industrial complexes in West Germany, which is economically so hard pressed and flooded with refugees.

---

[219]More precise figures and related information are available in Volker R. Berghahn's and Werner Abelshauser's contributions in Friess & Steiner (eds.) [1995] pp. 50f. and in Judt & Ciesla (eds.) [1996] pp. 33ff., 107ff.

[220]W. Hanle to M. Born, 13 March 1948, SBPK, manuscripts, Born papers 279 (reference to this letter by courtesy of Gerhard Rammer.)

In 1946 and 1947 there were demonstrations in many places in protest of the dismantling of factories of no military relevance.[221] Confidential reports like the one just quoted reveal a considerable potential for anti-Semitic and racist convictions also among scientists. The postwar tribulations and frustrations experienced in Giessen and elsewhere were only conducive to exacerbating the problem.

Fig. 14: Dismantled synthetic dye facilities of the *Badische Anilin- und Sodafabrik* (BASF) in Ludwigshafen, c. 1946. Source: BASF corporate archives, Ludwigshafen, Germany.

---

[221] For examples see Marshall [1980] pp. 667f., Benz (ed.) [1990] pp. 12–16, Overesch [1992] vol. 1. An article in the newspaper *Freiheit* dated 2 Dec. 1949 reports on a demonstration by factory workers in the Oppau plant of BASF dye works (company archive Ludwigshafen am Rhein). Particularly on the dismantling of the chemical industry see Raymond G. Stokes in Judt & Ciesla (eds.) [1996] p. 85.

7

SCAPEGOATING THE ARYAN PHYSICS MOVEMENT

As Alan Beyerchen, Mark Walker, and other historians of science have already pointed out, the retrograde Aryan physics movement of a few dozen active adherents was reduced after 1945 into a convenient scapegoat for all to thrash liberally without fear of reprisals.[222] One of its former standard-bearers, the Nobel laureate Philipp Lenard (1862–1947), was a broken man after 1945 and retired into obscurity in a village called Messelhausen. In a letter to Brüche he bitterly complained about "the loss of private property, a common fate in this predatory time" and the "rapid decline of all finer intellectual endeavors around the world," adding to his list of woes the insurmountable obstacles he was encountering in publishing a reedition of his book *Deutsche Physik*, and other works: "But alien conquerers have now deprived Gutenberg's fine invention completely of all nobler uses." Lenard does not figure in *Physikalische Blätter*, apart from as the subject of a modest necrology that leaves his political activities unmentioned (against the explicit advice of Pohl).[223] The somewhat younger champion of the movement, Johannes Stark (1874–1957) insisted on rendering his own account in 1947 "on the battles in physics during the Hitler period." Brüche even printed it, omitting only a final paragraph containing a few personal swipes. But he placed a rebuttal

[222]Brüche himself reports in *PB* **3** (1947) p. 167 (evidently proud about the effectiveness of the provocation), that Bruno Thüring's attorney filed a libel suit against the *PB* for describing him as "one of these sinister apparitions with turned-up collars, who put themselves forward in the coarsest possible manner" in *PB* **2** (1946) pp. 232f. Nothing further seems to have come of the suit. Cf. Hentschel (ed.) & Hentschel (ed. asst./trans.) [1996] pp. lxx–lxxvii on the Aryan physics movement (*Deutsche Physik*) and the references cited there; some pertinent documents are reprinted in this anthology. The historical transformation of this reactionary movement into a scapegoat after the war is discussed by Walker [1994] pp. 86f. and Eckert in Hoffmann & Walker (eds.) [2007].

[223]See the necrology by Wilhelm Orthmann (1901–1945), who in 1938 had compelled the chairmen of the DPG to urge all Jewish members to leave the society. The necrology of Lenard, *PB* **3** (1947) pp. 160f. reads: "In an effort to do justice to the great scientist [Lenard], our thoughts must reach back—bypassing the last few decades—to the beginning of the century." At the opening session of the convention of physicists on 5 Sep. 1947, von Laue worded his "distance" from Lenard as follows: "We cannot and do not intend to hide or excuse the mistakes of Lenard, the pseudo-politician; but as a physicist, he ranked among the great" (according to von Laue's letter to H. Pechel, 11 Nov. 1947, DPGA, no. 40048). A highly interesting, and to my knowledge, as yet unpublished typescript by Ramsauer on "Philipp Lenard as a scientist and a person" exists in two versions in BLM, box 2, folder 23.

by von Laue right after it, in which von Laue quoted extensively from arguments he had put forward in 1933 against electing Stark as a member of the Prussian Academy of Sciences.[224] Perhaps in order to underscore its scientific neutrality, the *Blätter* carried short articles and commentary in 1947 and 1948 on Stark's latest experiments on a supposed deflection of light in an inhomogenous electric field.[225]

Lesser proponents of Aryan physics were particularly convenient scapegoats for local faculties to unburden themselves of their Nazi pasts. For instance, Ferdinand Schmidt (1889–1960), a pupil of Philipp Lenard at Heidelberg, had been invited to substitute at the Stuttgart Polytechnic in November 1937 solely on the strength of his doctoral advisor's political influence. In April 1938 this assignment had been converted into a tenured professorship for fundamental physics. In August 1945 Schmidt was removed from office on the basis of the Liberation Law, and the tribunal decision of March 1948 ranked him among fellow travelers for lack of "distinguishing himself" politically—not to speak of academically. In a letter to the Ministry of Culture of Baden-Württemberg, the rector at Stuttgart thereupon argued that a reappointment of Schmidt was out of the question, recommending instead that two thirds of his former legal pension be granted him, "to the statutory regulation," because "after the collapse he had most diligently participated in the reconstruction [...] as far as his powers permitted. On the 15th of May of this yr. he collapsed in the street as a result of overwork. A heart condition prevents him from doing any more physical labor. His capital

[224]See Stark [1947] followed by von Laue [1947] and a notice by the editor, Brüche p. 288. Brüche also exposed as "'Aryan physics' in a new idiom" a right-wing organization in Peine, near Hildesheim, that had been newly founded under the innocent title: "Archive of the Independent Society for the Fostering of Young Science and Art": *PB* **5** (1949) issue no. 3, p. 152. Its editors were Klas Besser and Bruno Thüring, former editor of the publication issued by the Nazi student union, *Zeitschrift für die gesamte Naturwissenschaft*.

[225]See *PB* **3** (1947) pp. 162–163, **4** (1948) p. 71 and the notice by the editor, p. 168 at the end of issue no. 5, and Kleinert [2002]. On other attempts by Stark to submit more publications after 1945, see Lemmerich (ed.) [1998] pp. 474, 479–481 as well as von Laue's correspondence with R.W. Pohl 1952/53 in AMPG, div. III, rep. 50, no. 1908 and 593. A relatively lengthy necrology appeared in the *PB* **13** (1957), pp. 370f. by K. Kuhn. "Stark was, like his friend Lenard, a convinced National Socialist," psychologically tempering this statement with the follow-up: "He saw in Hitler a noble person who would save the German nation from doom and lead it to new heights. But when Hitler started the 2nd world war in 1939, Stark would have nothing more to do with Hitler." Andreas Kleinert has demonstrated that Stark only left the Nazi Party in 1943 because of local quarrels with the regional party leader.

is depleted."[226] This generous treatment by the polytechnic is quite astonishing, considering that Schmidt had originally been appointed "by the then Minister of Culture Morgenthaler against the wishes of the faculty at the instigation of the famous Nazi exponent of physics, Professor Lenard in Heidelberg." Indeed, "if physicists had ever heard of him, it was only as the proposed nominee who had been summarily rejected by two German universities."[227] Neither did Schmidt manage to make a name for himself while in office. Letters to the Ministry of Culture from mid-1945 and 1946 give particulars about his performance:[228]

> During his term in office at the Stuttgart Polytechnic, not a single publication issued from the institute that his predecessor, Professor Regener, had led to world fame. No research characterizes his directorship of the institute, just a series of unbelievable proceedings: two members of the staff he had hired had to be dismissed under grave circumstances, among them his son, whom he had engaged as managerial assistant despite his complete immaturity. His lectures remained well below the level worthy of a college. The intention of removing him from his chair already existed in 1943, within the framework of the planned reorganization of German physics, a measure which now appears of urgent necessity.

The justification given by the rector for nevertheless paying Ferdinand Schmidt two thirds of a regular pension was that, with the exception of his membership in the Nazi party (since 1 May 1933), he had not particularly incriminated himself politically. Besides, he had been in the civil service for two decades—this argument is a typical expression of the corporate spirit in German academia: all former colleagues must be embraced as best as possible; meanwhile the group of hitherto discriminated members must be held at arm's length and extreme cases shunned as black sheep.

The case of the Austrian professor of chemistry Jörn Lange (1903–1946) is an unusually prominent incidence of the impulse to defend the public image of scientists. In the final days of the war, Lange wanted to destroy the valuable electron microscope at the University of Vienna to prevent it from falling into the hands of the advancing Russian troops. When two of his colleagues tried

---

[226]Rector's office of the Stuttgart Polytechnic, signed L., to the Kultministerium [*sic*], 11 June 1948, carbon copy in UAS 57/411, personnel file on Ferdinand Schmidt.

[227]Ibid. Lenard's authoritarian manner of plying this political influence is documented by his letter to Reich Minister of Education Rust in Berlin, dated 22 May 1937, Bundesarchiv Koblenz no. 10 507 (alt R 21), vol. I, rep. 76, no. 507, with a copy also in UAS.

[228]Rector Grammel to Baden-Württembergische Kultministerium [*sic*], 5 Feb. 1946, p. 2, UAS 57/411; see also the letter to Landesverwaltung für Kultus, Erziehung und Kunst dated 26 July 1945.

to prevent him from doing so, he shot one (or both) in a wild shoot-out. The Austrian courts condemned him to death. Long after his sentence had been carried out the readers of the *Physikalische Blätter* were continuing to be fed with new documents presenting the various arguments submitted by the defense in the suit. Their client had purportedly acted in self-defense on one count and pled innocence of the second murder.[229] One comes away with the impression that the journal's intent was to bleach this perhaps unique case of politically motivated murder committed by a scientist, so as to be able to uphold the contentious thesis that no serious scientist could ascribe to the Nazi ideology so fanatically as to commit murder in the name of the cause.

Völkische science in general, and Aryan physics in particular, served as convenient scapegoats on which to shift the blame for the ideological embarrassments of the past. Following the logic of complicity constriction,[230] some ascerbic postwar texts seem to try to outbid each other with their jabs at the Aryan physics movement. Curt Wallach's article for *Deutsche Rundschau* is a particularly sarcastic example, calling the movement the "true fruit of the quagmire of Teutonic master-race insanity":[231]

> In the open country of Nazi derangement, that special department of the dozen-year madhouse, in which the champions of a new National Socialist worldview—ideologically vamped by Reichsleiter Rosenberg— proclaimed their brassy slogans, sprouted the bizarrest of outgrowths of that somewhat out-of-the-way field of "Völkische Science," modestly named "Aryan Physics."

Towards the end of his 14-page article full of excerpts, Wallach justified attempting to draw up a "medical report of Aryan Physics" to discuss that "truly unique spasm [...] as a psychopathology, just as we feel the urge to discuss the whole phenomenon of National Socialism psychologically. The purpose is not to find excuses for its actions and ravings, but to search out those dark and dangerous depths from which the dragons of Nazism hatched out and which must be recognized and elucidated to make what is incomprehensible humanly explicable."[232]

---

[229]This interesting case is outlined in *Neue PB* **2** (1946), issue no. 1, pp. 17f., *PB* **4** (1948) pp. 72f., 387–390. According to Deichmann [2001] p. 531, Lange had been a member of the Nazi party since 1933, earned his academic teaching degree in 1934 at the Univ. of Jena, and just been appointed supernumerary professor at the Univ. of Vienna in 1942.

[230]About this important concept of narrowing down the guilty to a few culprits (*Täterkreisverengung*), see Frei [1999] p. 405, Schüring [2006] pp. 365ff.

[231]See Wallach [1946]; on the following see Perron [1946] and Brüche [1946], quote on p. 233.

[232]Ibid., p. 138. This odd style was astonishingly well received. See Lise Meitner to Max von

The outcome of Wallach's reduction, still steeped in the language of the National Socialist era, is the sorry triad:

1. an "obstinately stubborn self-defensiveness against the rapidly changing times, which some are unwilling to follow along with because they are incapable of doing so."

2. "an emotional reaction, born of an unconscious sense of their own inadequacy," and

3. "a romantically skewed conservatism."[233]

The mathematician Oskar Perron wrote along the same lines in his article for *Die Neue Zeitung.* Ernst Brüche published extracts from both articles in 1946 in the recently renamed *Neue Physikalische Blätter,* adding excerpts from Wolfgang Finkelnburg's memorandum on "The fight against party physics."[234] We find the purpose of this monster show in Brüche's introduction to a reprinting of Stark's infamous commentary in the SS newpaper: "A reminder of this article in the 'Schwarze Korps' from 15 July 1937 is particularly worthy of note because it shows clearly the stance held by scientists, whom some over-eager tribunals and some foreign voices today would like to see as generally politically implicated." Stark's castigation of the overwhelming majority of professors at German universities and colleges as having disgracefully failed their country during the party's "time of struggle" [*Kampfzeit*] before 1933 is here taken, by implication, as proof of an unblemished political record extending to *beyond* the Nazis' rise to power. This "fight against party physics" waged by Ramsauer, Finkelnburg, and von Laue in the columns of the *Physikalische Blätter* is a form of "moral rearmament" by the physics community,[235] certifying before the public that during the Nazi regime they had stood on the right side, after all.

The long-time chairman of the DPG, Carl Ramsauer, presented similar evidence in his "History of the German Physical Society during the Hitler period," by concentrating on the efforts taken against the excesses of so-called *Deutsche*

Laue, 13 July 1947, quoted in Lemmerich (ed.) [1998] pp. 492f.: it was "very clear and in my opinion really reveals the main basis for the attitudes of people of the likes of Lenard."

[233] Wallach [1946], p. 141.

[234] For an English translation of this and the following document, Stark's commentary: "Science is politically bankrupt" on an anonymous anti-Semitic vituperation in the same issue of *Das Schwarze Korps,* see Hentschel (ed.) & Hentschel (ed. asst./trans.) [1996] docs. 110, 55f.

[235] As pointed out by Dreisigacker and Rechenberg [1994] p. 23.

Physik.[236] After 1945 he also portrayed the high-frequency expert and Knight of the Iron Cross Abraham Esau as a convenient black sheep, accusing him of "making peace with the National Socialist ministries instead of fighting a thankless battle for a higher cause." In early 1937 Esau had become head of the physics special division of the Reich Research Council (RFR) and was also appointed president of the bureau of standards (PTR) in 1939. While in power, Esau had moved the focus of German physical research toward military applications.[237] After the war, the Americans handed Esau over to the Dutch authorities to stand trial for his involvement in the plundering of research facilities of the Philips company. He was acquitted and expelled but later retried and convicted *in absentia*.[238] After 1945 Esau was nevertheless successful in regaining a foothold in Germany as a researcher, thanks to the support of the science policy-maker of North Rhine-Westphalia Leo Brandt (1908–1971): first as guest professor of short-wave technology at the Aachen Polytechnic, then in 1949 as head of the Institute of High-Frequency Engineering of the German Aeronautical Research Institute in Mülheim (Ruhr). If Max von Laue had not intervened, he would have even been awarded a medal of distinction from the Federal Republic at Leo Brandt's nomination in 1954. Von Laue pointed out that, because Esau "had presented himself at all official functions during the Hitler period as the chief representative of National Socialism," his acquittal in the Netherlands had "astonished" many German physicists. It would be even less appropriate to confer an award on him now as well.[239] Esau's prewar policy of militarizing research went beyond what his colleagues within the physics community were willing to close an eye to out of solidarity.

The same scapegoat strategy was used by physicists within the Soviet zone. Robert Rompe in East Berlin, for instance, decried in 1947 the "misrepresentations of the exact sciences" during the Nazi period and denounced Lenard and Stark as "men who had participated actively in the endeavors of the NSDAP," because

---

[236] See Hoffmann [2002] and Hentschel (ed.) & Hentschel (ed. asst./trans.) [1996] docs. 86, 88, 90–93, 110.

[237] Ramsauer [1949] p. 78. Hoffmann & Stutz [2003] pp. 136, 161f., 168ff. interpret more plausibly Esau's policy of attempting to reconcile the interests of the army, major industry, and the state with physics. The letter by a secret service official, Helmut J. Fischer to E. Brüche, 27. Nov. 1948, BLM, box 104, folder 261, gives a perhaps better informed assessment of Esau's strengths and weaknesses.

[238] Rusinek [1998] p. 44 supposes this conviction was necessary "because the Dutch needed a legal claim for Philips to file for damages from the Federal Republic." See also Grüttner [2004] p. 45.

[239] See Rusinek [1998] p. 44.

they had employed "the aggressive methods fostered there, of which political anti-Semitism formed a part."[240] The argument then takes a strange turn:[241]

> One should probably grant Stark and Lenard a certain degree of honesty. The methods of modern physics were obviously really not evident to them. [...] Although the era of "Aryan physics" was in clear decline from 1937/38 on and progressive scientists waged an effective battle against it behind the scenes, many classes of promising young physicists were prevented from devoting their full energy to this field. The number of men publicly putting their academic standing at risk in confronting the proponents of the medieval interpretation of science promulgated by the Nazis was not large. [... Max von Laue and Walter Weizel are then named in particular.] The more highly we must rate the courage of these men for taking this stance. It shows that opposition to the machinations of the Nazis was very certainly possible, leading even to success, if those under attack manage to unite.

There is a pattern here: "reducing the circle of culprits" through repudiation of a few strongly implicated members of a defined group, in order to immunize the community as a whole. This pattern also extends beyond the physics community (where Stark, Lenard, and Wilhelm Müller figured as the main scapegoats) to other scientific institutions. The Kaiser Wilhelm/Max Planck Society[242] also projected a public image of "probity among our circle of colleagues" and "purity among our ranks" by announcing "corrections" in the composition of the governing board. In 1945, the party members Wolfgang Finkelnburg and Max Steenbeck were replaced by Ferdinand Trendelenburg (1896–1973) and Josef Krönert (born 1891) on a temporary basis.[243] It is another characteristic of this pattern that the finger-pointing comes to a sudden stop, once the hot denazification phase has begun to cool. The actors assume a "draw-the-line mentality" (Norbert Frei) and allow the misdeeds to sink into the obscurity of history. When in 1951 Max von Laue offered the editor of the *Physikalische Blätter* a stenographed transcript of

---

[240]Rompe [1947] p. 136.

[241]C. Ramsauer [1947] p. 114. The introduction in *Neue PB* **2** (1946) p. 2 assumes a similar note: "it is no small consolation that with the downfall of the despotic regime, the building grounds for free thought structures are being cleared again."

[242]See, e.g., Schüring in: vom Bruch & Kaderas (eds.) [2002] p. 462 and Sachse [2002] on the Verschuer case.

[243]See DPGA, no. 40025. Cf. Pohl to Regener, 24 July 1946 (ibid., no. 40027): "It is self-evident that only politically unimplicated persons can sit on the board. It also seems of secondary importance to me whether or not political aspects are considered in the selection of members in the individual zones."

Johannes Stark's speech delivered at the physicists' convention in Würzburg on 18 September 1933, Brüche politely declined:[244]

> I have my doubts, however, about whether one ought to publish it in the foreseeable future [...], not just because there actually should be an end to the "thousand-year Reich" along with its entire aftermath, but particularly because the Physikalische Blätter has gradually been evolving in a new direction.

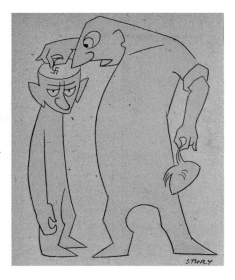

Fig. 15: Cover of the satirical magazine *Das Wespennest*, no. 9, 7 Oct. 1948. Cartoon by Stury. Source: Haus der Geschichte, Bonn, EB no. 1994/05/0290.

[244] E. Brüche to M. von Laue, 15 May 1951, AMPG, div. III, rep. 50, no. 371. This transcript is particularly interesting for the passages omitted from its published version in *Zeitschrift für technische Physik* 14th series, no. 11 (1933), pp. 433–435. As another contemporary, Friedrich Hund, also told me, Stark actually closed the speech with the following threat to uncooperative colleagues: "and if you are unwilling, I will use force!": Hentschel (ed.) & Hentschel (ed. asst./trans.) [1996] pp. 71–76 and von Laue [1947].

8

FORGETTING

The following discussion of this important complex of postwar mentality distinguishes unconscious repression of the past from intentional suppression and concealment of facts.

(a) Amnesia and unconscious repression

Like many emigrés, Lise Meitner was deeply troubled that so few of her former colleagues were able to admit their own mistakes and crimes even to themselves after the reign of terror in Germany had come to an end. As her letters to Otto Hahn illustrate, she was particularly disappointed that they were unable to see any moral weakness or failure in their behavior during the Nazi period.[245] Meitner succinctly reiterates this point in another letter to James Franck in 1947, after Hahn had received the Nobel prize in Sweden: "His interviews all sound the same. Just forget the past and emphasize the injustice happening in Germany."[246] This forgetting easily led to embarrassing misunderstandings of grotesque proportions, as Meitner recalled in this letter. She was trying to understand Hahn's and von Laue's mentality, which, she surmised,

> they probably share in common with many decent Germans. Hahn came here possessed by two thoughts. First, that the Germans were now suffering severe injustice particularly from the Americans, and he had written me a couple of weeks before his arrival that the Americans were doing the same things in Germany that the Germans had done in Poland and Russia. Even though I replied to him, in writing, that it really was impossible that he could seriously think that and that he cannot have forgotten that in Poland 2 million people had been murdered and practically all cultural artefacts had been systematically destroyed, he still repeated those very words when he was with Stern and me; which upset Stern very much. He simply did not respond to counterarguments, he is suppressing the past with all his might,

---

[245] Her accusing letters to her friend and former colleague, O. Hahn, have already been cited so often that I may dispense with quoting them. See, e.g., Hentschel (ed.) & Hentschel (ed. asst./trans.) [1996], docs. 108 and 120 for her letters from 27 June 1945 and 6 June 1948 in English translation.

[246] Meitner to J. Franck, 16 Jan. 1946, RLUC, box 5, folder 5. Cf. Deichmann [2001] p. 458. Bröckling [1993] gives a more general portrayal of the forgetting and denial in postwar Germany.

even though he always had really hated and disdained the Nazis. But since his second main motive is to return Germany to international repute and he is neither a strong character nor a very contemplative person, he simply denies what has happened or makes light of it. During the first interview he gave for Sweden he did not mention a single word about the past or a wish that Germany would become a constitutional state again and thus develop into an equal partner in Europe [...].

Laue is better capable of reflection than Hahn, of course, and does respond to whatever issues he is directly confronted with; but his basic attitude is not very different from Hahn's. Maybe our generation is too old to look things straight in the eye and no longer has the energy to wage true battle against the confused ideas that have prevailed for well over a hundred years and have only found particularly gruesome expression in Nazism. I, at least, find it disturbing for Germany's future and for the world's future.

There is a less biased assessment in Meitner's letter to von Laue from June 1946, in which she notes the phenomenon of forgetting across the board:[247]

if I weren't such an inveterate opponent of pessimism, I would lose all faith in any practicable solution for the future. It all looks like one big general confusion. One side seems to have forgotten what they promised, the other side to have forgotten what has happened and what has been condoned.

The former official of the Prussian Ministry and deputy main prosecutor of the US at Nuremberg, Robert Kempner (1899–1993), significantly referred to "repression acrobats." The chemist Richard Castleman Evans (born 1916), who at the behest of the Allies traveled extensively throughout Germany and Austria as an inspector in 1946 and had spoken with a large number of scientists in their laboratories, gained the following general impression:[248]

Few scientists have the inclination to speak about the war; they regard it as an unfortunate interruption to their normal employment and think that it now should be completely forgotten. No one has any sense of guilt about the war. "And then the war came," is the uniform statement that ends almost all reports about studies and research in 1939. No scientist had ever been a Nazi except for those who cannot hide the fact owing to their suspension from office. And they, too, have long since found means to assuage their consciences.

---

[247] Published by Jost Lemmerich (ed.) [1998] pp. 451–453, quote on p. 452.

[248] Evans [1947], quote on p. 13. According to the 10th edn. of *American Men & Women of Science*, Evans was research chemist under National Defense Contract 1942–46, changing subsequently to the Ballistics Lab, Aberdeen 1946/47, thereafter to the Catalyst Research Corporation, Baltimore. Kempner had the reputation of being a forceful interrogator. See Frei [1999] pp. 103f., 149ff.

Margarete and Alexander Mitscherlich have argued—in my view, convincingly,—
that this repression was unconscious. It obscured injured feelings of self-worth and
ultimately an "inability to mourn."[249] The much cited *Vergangenheits"bewältigung,"*
or coming to terms with the past, more than coincidentally resembles the word
überwältigen, or overcoming, that is, with the intent of getting the past over and
done with, by sheer collective will.[250] Individuals awoke quite late from this
stunned traumatic state. On 20 July 1974, the 30th anniversary of Stauffenberg's
assassination attempt on Hitler, Carl Friedrich von Weizsäcker delivered a speech
in the courtyard of the former Army headquarters in Berlin, the scene of the
executions of the perpetrators. He called on Germans to let "confession [now]
follow this curative slumber":[251]

> It would be healthy for us, if it were not shameful to admit that we followed
> Hitler, that each of us, perhaps to different degrees and at different stages,
> were members of a National Socialist nation. If we could shed the guilt
> complex, perhaps we can emerge from the repressed depths and allow
> ourselves belatedly to mourn.

In the chapter on self-justification and the guilt issue (on p. 116) we shall see
that psychological barriers were not the only obstacles to open admission of
guilt. Strategic political reasons were also at play. As Carola Sachse and Michael
Schüring have demonstrated with respect to members of the Kaiser Wilhelm
Society, "repression" of what had happened or psychological banishment from
the collective conscience had a corollary of "repulsion." Strongly implicated
individuals within a given scientific community were dismissed from office and
deactivated in order to take no longer presentable elements out of the line of fire
(in a sense, to shorten the front line).[252]

---

[249] See Mitscherlich & Mitscherlich [1967] and Arendt [1993] p. 24; cf. Szabó [2000] pp. 508f.
for criticism of this interpretation as unconscious repression. There the refusal to remember is
conceived as an active process. Bröckling [1993] p. 55 refers to a collective "unwillingness to grieve"
and a continuation of defense strategies diagnosed in his book. He also offers a psychological
interpretation of the West German reaction to Mitscherlich's "soiling of the nest." Brumlik [2001]
extends this further, referring to a refusal to feel compassion ("Einfühlungsverweigerung").

[250] More recent analyses therefore prefer to use the term policy (*Vergangenheits"politik"*): see, e.g.,
Frei [1999] p. 14, Weisbrod [2002], Schüring [2006] p. 7.

[251] Wein [1996] p. 378. Max Planck's son Erwin (1893–1945), a former officer and state secretary,
working in industry since 1936, was among those arrested after the failed coup and convicted to
death by Freisler's *Volksgerichtshof.* Despite Max Planck's efforts to intervene, the death sentence
was executed in late January 1945.

[252] See here the chapter on scapegoating Aryan physics, p. 91, and Sachse [2002], Schüring [2006]
pp. 274f.

MG/PS/G/9a

MILITARY GOVERNMENT OF GERMANY
Fragebogen

WARNING: Read the entire Fragebogen carefully before you start to fill it out. The English language will prevail if discrepancies exist between it and the German translation. Answers must be typewritten or printed clearly in block letters. Every question must be answered precisely and conscientiously and no space is to be left blank. If a question is to be answered by either "yes" or "no", print the word "yes" or "no" in the appropriate space. If the question is inapplicable, so indicate by some appropriate word or phrase such as "none" or "not applicable". Add supplementary sheets if there is not enough space in the questionnaire. Omissions or false or incomplete statements are offenses against Military Government and will result in prosecution and punishment.

WARNUNG: Vor Beantwortung ist der gesamte Fragebogen sorgfältig durchzulesen. In Zweifelsfällen ist die englische Fassung maßgebend. Die Antworten müssen mit der Schreibmaschine oder in klaren Blockbuchstaben geschrieben werden. Jede Frage ist genau und gewissenhaft zu beantworten und keine Frage darf unbeantwortet gelassen werden. Das Wort „ja" oder „nein" ist an der jeweilig vorgesehenen Stelle unbedingt einzusetzen. Falls die Frage durch „Ja" oder „Nein" nicht zu beantworten ist, so ist eine entsprechende Antwort, wie z. B. „keine" oder „nicht betreffend" zu geben. In Ermangelung von ausreichendem Platz in dem Fragebogen können Bogen angeheftet werden. Auslassungen sowie falsche oder unvollständige Angaben stellen Vergehen gegen die Verordnungen der Militärregierung dar und werden dementsprechend geahndet.

A. PERSONAL / A. Persönliche Angaben

1. List position for which you are under consideration (include agency or firm). — 2. Name (Surname). (Fore Names.) — 3. Other names which you have used or by which you have been known. — 4. Date of birth. — 5. Place of birth. — 6. Height. — 7. Weight. — 8. Color of hair. — 9. Color of eyes. — 10. Scars, marks or deformities. — 11. Present address (City, street and house number). — 12. Permanent residence (City, street and house number). — 13. Identity card type and Number. — 14. Wehrpaß No. — 15. Passport No. — 16. Citizenship. — 17. If a naturalized citizen, give date and place of naturalization. — 18. List any titles or nobility ever held by you or your wife or by the parents or grandparents of either of you. — 19. Religion. — 20. With what church are you affiliated? — 21. Have you ever severed your connection with any church, officially or unofficially? — 22. If so, give particulars and reason. — 23. What religious preference did you give in the census of 1939? — 24. List any crimes of which you have been convicted, giving dates, locations and nature of the crimes. —

1. Für Sie in Frage kommende Stellung: ....Spruchkammer..............

2. Name .... 3. Andere von Ihnen benutzte Namen
    Zu-(Familien-)name   Vor-(Tauf-)name

oder solche, unter welchen Sie bekannt sind ....

4. Geburtsdatum .... 5. Geburtsort ....

6. Größe .... 7. Gewicht .... 8. Haarfarbe .... 9. Farbe der Augen ....

10. Narben, Geburtsmale oder Entstellungen ....

11. Gegenwärtige Anschrift ....
                            (Stadt, Straße und Hausnummer)

12. Ständiger Wohnsitz ....
                            (Stadt, Straße und Hausnummer)

13. Art der Ausweiskarte .... Nr. .... 14. Wehrpaß-Nr. .... 15. Reisepaß-Nr. ....

16. Staatsangehörigkeit .... 17. Falls naturalisierter Bürger, geben Sie Datum und Einbürgerungsort

an ....

18. Aufzählung aller Ihrerseits oder seitens Ihrer Ehefrau oder Ihrer beiden Großeltern innegehabten Adelstitel ....

19. Religion .... 20. Welcher Kirche gehören Sie an? .... 21. Haben Sie je offiziell oder inoffiziell Ihre Verbindung mit einer Kirche aufgelöst? .... 22. Falls ja, geben Sie Einzelheiten und Gründe an .... 23. Welche Religionsangehörigkeit haben Sie bei der Volkszählung 1939 angegeben? .... 24. Führen Sie alle Vergehen, Übertretungen oder Verbrechen an, für welche Sie je verurteilt worden sind, mit Angaben des Datums, des Orts und der Art ....

B. SECONDARY AND HIGHER EDUCATION / B. Grundschul- und höhere Bildung

Name & Type of School (If a special Nazi school or military academy, so specify) / Name und Art der Schule (Im Fall einer besonderen NS oder Militärakademie geben Sie dies an)	Location / Ort	Dates of Attendance / Wann besucht?	Certificate Diploma or Degree / Zeugnis, Diplom oder akademischer Grad	Did Abitur permit University matriculation? / Berechtigt Abitur oder Re-fe-xcognitis zur Universität-smatrikulation?	Date / Datum

25. List any German University Student Corps to which you have ever belonged. — 26. List (giving location and date) any Napola, Adolf Hitler School, Nazi Leaders College or military academy in which you have ever been a teacher. — 27. Have your children ever attended any of such schools? Which ones, where and when? — 28. List (giving location and date) any school in which you have ever been a Vertrauenslehrer (formerly Jugendwalter).

25. Welchen deutschen Universitäts-Studentenburschenschaften haben Sie je angehört? ....

26. In welchen Napola, Adolf-Hitler-, NS-Führerschulen oder Militärakademien waren Sie Lehrer? Anzugeben mit genauer Orts- und Zeitbestimmung ....

27. Haben Ihre Kinder eine der obengenannten Schulen besucht? .... Welche, wo und wann? ....

28. Führen Sie (mit Orts- und Zeitbestimmung) alle Schulen an, in welchen Sie je Vertrauenslehrer (vormalig Jugendwalter) waren ....

C. PROFESSIONAL OR TRADE EXAMINATIONS / C. Berufs- oder Handwerksprüfungen

Name of Examination / Name der Prüfung	Place Taken / Ort	Result / Resultat	Date / Datum

C. Brügel & Sohn Ansbach

Fig. 16: First page of a questionnaire issued by the US Military Government of Germany about membership and offices held in the Nazi party and its affiliated organizations, 1945 (source, e.g., BLM, box 123, folder 321).

(b) Concealment and dissimilation

This section will discuss documentable cases of falsely filled-out questionnaires. All university faculties were affected but the chaotic state of the verification offices meant that many incorrect responses were never exposed.[253] It is highly interesting to observe the evolution of Ernst Brüche's responses to the question regarding his "official capacities" and see what exactly he was willing to divulge and when, in the various drafts for a series of questionnaires issued by the Office of Military Government.[254] His diary also shows how much he agonized over certain touchy points:[255]

> Is "security agent" considered a punishable relationship with the SD? I very vaguely remember that sometime right at the end of the war some sort of reorganization had been carried out making security agents from then on subordinate to the SD. But I don't remember exactly anymore. But I have to say that I was a security agent in Schönberg. I hadn't thought of it before, now I have to indicate it on the forms that have to be submitted right now. Will they believe me that I didn't do any harm? [...] But they might simply hold against me that I had not mentioned this activity on the old questionnaire. How is one to remember such things! I don't know; I don't quite trust the Americans on this point. They are far too formally minded and then, what do they know about German circumstances?

---

[253] At the Univ. of Göttingen, for instance, Richard Becker, Arnold Eucken, and Hans König gave wrong or incomplete information in response to questions regarding politics, as revealed in Rammer [2004] on the basis of the card index at the former Berlin Document Center. Wrong information by students was even more difficult to check, see Rammer [2004] pp. 214f. Inadvertent self-incriminations also occurred, as indicated, e.g., in Sommerfeld's character reference of Ernst von Angerer, 10 Feb. 1946: "His dismissal was received with general astonishment and can only have arisen from misunderstood or perhaps unclear formulations he made in his questionnaire." www.lrzmuenchen/~Sommerfeld/gif100/05159_01.gif Such questionnaires issued by the Allied occupiers are discussed and evaluated in Tent (ed.) [1998] pp. 11, 176f. Bird [1978] pp. 147ff., reports that the sheer number of questionnaires filled out by applicant students prevented proper individual political evaluations as required prior to admission. In actual practice they were ignored. According to Brüdermann [1997] p. 104, from out of a total of 2.1 million checked questionnaires in the British zone, 2,345 of the respondents (i.e., 0.1%) were forced to bear the consequences of giving false information by the end of 1947. The subset of persons deciding to change their identities using false papers is discussed by Frei in König et al. (eds.) [1997] pp. 207f.

[254] See Brüche's denazification questionnaires or drafts dated 7 June 1945, 14 Sep. 1945, 12 and 13 Nov. 1945, 29 May 1946 and 4 Dec. 1946, all in Annex 1 at BLM, box 123, folder 321.

[255] E. Brüche, diary no. VI, 11 May 1946 (BLM, box 7); ibid., 11 May 1946 on the dismissal of a Dr. Fuchs for having neglected to mention in his questionnaire financial contributions he had made to the SS.

This fear that "the Americans" would misunderstand their situation reveals the deep distrust of the occupiers. But among physicists I have not encountered quite so extreme a case as that of the rector at Erlangen, Theodor Süß. He not only hushed up his Nazi memberships right after the war but even saw to it that "during the remainder of his term as rector a number of professors received calls to the University of Erlangen who had previously been dismissed at other universities for political reasons. It is significant that the denazification committee installed by Rector Süß was locally nicknamed 'Club for the Salvation of the Shipwrecked'."[256]

Countless instances of surreptitious cleansing of documentary evidence also belong in this category. It was most easily carried out by people in leadership positions after 1945 (e.g., as dean or rector) who had free access to staffing files.[257] In filling out his questionnaire, the theoretical physicist Friedrich Möglich (1902–1957) made no mention of his membership in the Nazi party between 1932 and 1938.[258] Instead he referred to an attached exonerating opinion that was intended to make him acceptable for appointment as professor and director of the Institute of Theoretical Physics at the Humboldt University in Berlin and as manager of the Physics Section of the Institute of Medicine and Biology in Buch near Berlin, subordinate to the Soviet Military Administration until July 1947. His less privileged contemporaries were left the option of a "*U-Boot* career," as life under a false name was euphemistically called in the young Federal Republic of Germany. According to an official estimate from 1954, around 60,000 persons were living under false identities at this time in the FRG alone, with forged or substitute paperwork based on false facts. The impostor Schwerte/Schneider managed to attain the post of rector at the Technical University in Aachen before his Nazi past caught up with him. Bernd-A. Rusinek writes about the career of this "splits actor" after 1945:[259] "Going underground was relatively easily done. Although universities were at all times notorious for their climate of gossip and

---

[256] *Die Neue Zeitung*, 3rd ser., no. 10, 3 Feb. 1947, p. 5: "76 Entlassungen an der Erlanger Universität."

[257] Only individual cases are documentable, but numerous coincidences speak volumes by implication. The personnel files for Walter Tollmien for the period 1937–55 at the Dean's Office and for Richard Becker for the period preceding 1948 at the Rector's Office are completely empty (Becker happened to hold the office of Prorektor at the Univ. of Göttingen in 1948). By contrast, parallel files at other administrative offices are bursting at the seams. (I am indebted to Dr. Rammer for pointing out these two cases.)

[258] Dieter Hoffmann drew my attention to the case of Friedrich Möglich, expelled from the Nazi party for having committed a "racial disgrace" (*Rassenschande*). See Hoffmann & Walker in Walker (ed.) [2003].

[259] Rusinek [1998] pp. 34 and 20 on the official estimate of persons living under false identities.

intrigue, after 1945 the equivalent of involvement often shielded Germans from enduring the consequences of information about a university teacher's brown past. A mentality of emergency mutual assistance dominated."

At first, many tried to look at the humorous side of the obligatory questionnaires. Max von Laue wrote his son Theo:[260] "In the Middle Ages there was a period of round arches [*Rundbogen*] and one of pointed arches [*Spitzbogen*]. We are living in a period of the *Fragebogen* [questionnaires]." But when new questionnaires kept on coming, the novelty soon wore off: "More 'denazification' is going on here. My colleagues and I are now supposed to fill out our fourth questionnaire, a monster of 12 pages and with 133 questions! We declared that we are refusing to fill it out. The thing is beginning to get humiliating." To one question posed in the Military Government's questionnaire of 1948 about when and where other questionnaires had been filled out, Fritz Houtermans responded: "Göttingen 1945–47 continuously."[261] The national conservative writer Ernst von Salomon (1902–1972) carried this joke to the extreme. In his book, *Der Fragebogen*, first published in 1951 and still in print until the end of the 1970s, selling a total of about 200,000 copies, he ironically asked: "How else should I interpret the structure of the questionnaire than as a modern attempt to move me to examine my conscience?" and proceeded to fill 670 paperback pages with lengthy, and often long-winded stories in response to each question. It thoroughly exposed the inappropriateness and naïveté of many of the questions.[262] The secret to success for this not particularly literary work was that it expressed what many compatriots were thinking. The author found fitting words for the distrust he felt, indeed a growing "anger about the Americans": "What chokes every attempt to converse with Americans is their horrendous self-righteousness. They prove every moment that they don't know a thing but they act as if they always knew better. With sure

---

[260] On the following see M. von Laue to Theo, 18 April and 5 Sep. 1946, AMPG, div. III, rep. 50, suppl. 7/7, sheets 6, 46. Cf. sheet 61 (27 Nov. 1946: "The fact that I answered exactly the same question once already in order to get my pension is completely ignored. Well—I'll just wait and see how things turn out."). Local Berliners jokingly added the following 134th question to the questionnaire: "Have you ever owned a *Kanonenofen* [round iron oven]? That would certainly be a sign of militarism!" (sheet 28, 16 July 1946). For a collection of various drafts concerning a single individual on five different occasions between mid-1945 and the end of 1946, see also footnote 254 on p. 103 above.

[261] Houtermans in the questionnaire by the Military Government, 31 March 1948 (UAG), cited from Rammer [2004] p. 128. A seven-page questionnaire by the US Military Government of Germany with 131 questions is accessible at: www.hbg.ka.-bw.schule.de/publikat/ka45/divers/bogen01...07.gif

[262] See the spoof on a conscientious respondent: von Salomon [1951].

ease they inexorably choose the worst among all possible measures. And the
worst thing is, they never stop working against their own interests!" To which his
girlfriend in the book replies: "Just like the Nazis! [...] You're talking as if you were
corrupted by the Nazi propaganda against the Americans!" Ernst von Salomon's
response was: "Perhaps! What's certain is that the Americans are corrupted by
their propaganda against us!"[263]

For the sake of balanced reporting let me point out here that the Allies
themselves were not above breaking their self-imposed restrictions and obscuring, destroying, and keeping quiet about incriminating documents when it was a
matter of winning over the valuable skills of certain specialists. Under the code
names "overcast" and "paperclip" 1,500 scientists and technicians were allowed
into the United States against immigration regulations forbidding entry to former
members of the Nazi party and its affiliated organizations, not to speak of naturalization. Some of these specialists were very seriously incriminated. Arthur
Rudolph (1906–1996), for example, was responsible for designing the V-2 rocket
in the Dora underground concentration camp. In the East the Soviet Military
Administration made sure that over 50 percent of the former Nazi employees at
the PTR were immediately reemployed, in flagrant breach of the Ordinance for
the Purging of the Administration of Nazi Elements from 23 July 1945. They were
reengaged at the German Bureau of Weights and Measures built in East Berlin,
the new parallel institution in the Soviet sector to the Federal Republic's revamped
PTR.[264] In the case of the British, by March 1946, about 200 such scientists and
engineers were already reemployed under the codename "operation dustbin."[265]
The competition among the various Allied occupiers for highly qualified scientists
was an important reason for the sometimes astonishingly smooth reintegration
of politically implicated physicists. Some used this inter-Ally rivalry as effective
leverage in their employment negotiations.[266]

A somewhat subtler form of dissimilation is publication of "revised" texts
dating back to before 1945 without mentioning subsequent omissions of all politically suspect passages. Carl Ramsauer's reprinting of his "Kampf" against Aryan
physics is one example. All questionable statements about the "justifiable battle

---

[263] E. von Salomon [1951], on p. 535 of the edition from 1961. The book is cited, e.g., by Hückel
[1975] p. 162.

[264] Peltzer [1995] pp. 76–82 discusses the neglect to denazify PTR members. On American
"intellectual reparations": ibid., pp. 9, 23, 30ff. as well as Bower [1987], Gimbel [1990] pp. 37ff.

[265] See Farquharson [1997] p. 36 and Gimbel [1990] p. 17.

[266] Beyler [1994] pp. 469ff. provides the example of Pascual Jordan; further examples are discussed
in Judt & Ciesla (eds.) [1996] pp. 37, 62ff., 93ff.

against the Jew Einstein and against the outgrowths of his speculative physics" were silently struck out along with any mentions about his close cooperation with the Reich Ministry of Aviation, among other centers of National Socialist power.[267] Pascual Jordan's book *Die Physik und das Geheimnis des organischen Lebens* from 1941 was reprinted after the war without the slightest mention of any revisions, even though passages about the "purpose and importance of physical research [...] through its role as a technical and military instrument of power" among other politically steeped remarks had simply vanished. The young physicist Ursula M. Martius (born 1921) pointed out these discrepancies in 1947 in an article in the leftist liberal paper, bitterly complaining that such discredited persons remained in "positions in which they exerted influence on others and retained decision-making power."[268] This isolated protest had no effect at all. Other textual manipulations are similarly notorious. Helmut Albrecht's later close examination of Planck's report about his audience with Hitler reveals that it underwent significant changes before it appeared in print in the *Physikalische Blätter* in 1947.[269]

This leads us to hasty reinterpretations of past attitudes: for instance the actual intentions and goals in designing the uranium machine as agreed post factum by German nuclear physicists during their internment in Farm Hall.[270] The vacillating attitude of the astronomer at Göttingen and Hamburg Otto Heckmann (1901–1983) for or against—or rather for—the theory of relativity, after all, depending on how it suited the political situation, is another case which I analysed together with Monika Renneberg. We could join Mitchel Ash in calling it more neutrally a "restructuring of the resource constellation." Others will rather be reminded of the turncoats after 1989 or other standard-bearers allowing their banners to fly according to which ever wind happens to be blowing. In any event, these are

[267] See Ramsauer in *PB* **3** (1947) pp. 113–115; cf. Simonsohn [1992] and Hoffmann [2002] pp. 284–287. Arnold Sommerfeld's whitewash certificate for Ramsauer (22 Mar. 1946) focuses exclusively on his battle against Aryan physics: see http://www.lrz-muenchen/~Sommerfeld/gif100/05181_01.gif

[268] According to Martius [1947] p. 100 such persons "had lost every right to be an educator today. [...] the only assistants and staff members [they will] engage are people who pose them no threat, whom they either know from before, share the same attitudes or whom they have under their control through some sort of incriminating involvement." On Martius's case see Gerhard Rammer's contribution in Hoffmann & Walker (eds.) [2005].

[269] See Planck: *PB* **3** (1947), p. 143 and Helmut Albrecht in Albrecht (ed.) [1993] pp. 50f. The original draft apparently broadly alluded to Eastern Jews as worthless. Brüche's letter to Sommerfeld from 27 May 1947 documents that he was soliciting other similar articles—that is, ones putting physicists in a better light.

[270] See the transcripts of secretly recorded conversations of the internees: Frank (ed.) [1993].

all conscious maneuvers to conform and dissimilate. In 1945 this tendency fell within the context of a much more widespread and powerful wave of repressed responses, that "great alliance of silence and minimization of past evils."[271]   A correspondent of Brüche's referred in 1947 to "this world of the weak in spirit, in this time when worldviews can be had for a cigarette."[272] The few younger physicists (such as Kurt Hohenemser or Ursula Martius), who insisted on personal consequences for incriminated individuals, were ostracized as "soilers of the nest" and are not representative of the mentality of postwar physicists. Martius spent the last few months of the Nazi regime in a concentration camp with her father and Jewish mother. Not long afterwards, in September 1947 at the annual convention of the German Physical Society in Göttingen, she saw "the people who still appear to me in my nightmares [...] sitting there alive and unchanged in the front rows. Unchanged, if you don't consider the simple blue suit instead of the uniform and the missing party badge a 'change'."[273] Thus it is not surprising that she, like Hohenemser and other persecuted persons during the Nazi era, decided to emigrate after 1945—in her case to Canada, where she became an internationally recognized materials scientist and was appointed in 1967 the first female full professor at the University of Toronto.

[271]Walker [1990], [1993] discusses the postwar legend of political resistance by German nuclear physicists against the Nazi regime. On Heckmann see Hentschel & Renneberg [1995]. Cf. also Ash [2001] and Hoffmann [2002] quote on p. 289.

[272]Leo Prior to E. Brüche, 29 Aug. 1947, BLM, box 2, folder 24.

[273]Martius [1947], p. 99 (she specifically named Stuart, Schumann, Jordan, Schober, and H. Kneser, among others). This convention and the exchange between Martius and von Laue is discussed by Gerhard Rammer in Hoffmann & Walker (eds.) [2005]. For Martius's later walk of life as an "archaeometrist promoting world peace" see http://collections.ic.gc.ca/heirloom_series/volume6/284-285.htm

Fig. 17: American poster depicting atrocities committed in concentration camps, 1945. The aim of creating a sense of guilt among the German public failed (see here p. 171). Klessmann [1991] p. 308; see also Brink [1998] pp. 72f. on the instrumentalization of concentration-camp photos.

9

SHAME, LISTLESSNESS, AND LETHARGY

"They do not understand the situation and their own responsibility and [they] asked him whether Carnegie or Rockefeller Foundation could not finance their institutes! They think they lost the war because the scientists were not properly made use of!" Thus reads Max Born's diary entry about the impressions the emigré aerodynamicist Theodore von Kármán (1881–1963) brought back with him from a brief stay in Göttingen in the spring of 1946.[274]

One has to search long and hard to find statements by physicists openly confessing shame about the countless crimes committed against humanity during the Nazi period. Max Planck to James Franck, for instance, as early as 1935 in reply to an invitation to Denmark: "No, I cannot travel abroad. During my earlier trips I felt like a representative of German science and was proud of it—now I would have to hide my face in shame."[275] Planck was alluding here to the wave of anti-Semitic dismissals and perhaps also to the foreign policy of the new men in power, not, however, to the later policy of extermination. In 1943 he confessed to Lise Meitner during a visit in Stockholm: "Terrible things should happen to us, we have done terrible things."[276] In this instance, however, it was less a matter of shame than fear of reprisals after the lost war (not that he felt he would be personally punished. It was rather consternation about the serious consequences for physics as a whole in Germany).

The scarcity of such statements does not imply that a sense of shame really had been so rare. I would think it unlikely, considering the barrage of horrendous news reports and pictures from the extermination camps in all the media. The shock among those who in fact had not known what had been going on in those camps until the spring of 1945 must have been so great and the shame

---

[274]Original English. Quoted from M. Born's diary-like notes "Journey to Russia" under the date 15 June 1945 (copy in the Polányi papers, RLUC, box 8, folder 17). Polányi had met Kármán briefly during a stopover in Paris. Brüche's diary reveals how very apt this impression was; cf. his manuscript "Unsere neue Aufgabe" (cited here in foonote 64 on p. 28).

[275]M. Planck to J. Franck, Franck papers, RLUC, cited from Lemmerich [2002] p. 131.

[276]Lise Meitner in a letter to Max von Laue, 22 Feb. 1946, Franck papers, RLUC, box 5, folder 5; cf. also Lemmerich (ed.) [1998] p. 452. Elsewhere Meitner continues: "(it is genuine Planck for him to have said 'we' and 'us,' and I could have kissed his hand for that)."

so overwhelming that it rarely found verbal expression.[277] On 22 April 1945 a still somewhat incredulous Ernst Brüche wrote in his diary: "If everything that the Allied broadcaster is suggesting about the concentration camp in Buchenwald really is true, that in March 17,000 people died there of hunger and sickness, then that officer was right, who after falling into captivity and registering it with his own eyes said, he had now also lost his honor. If what is being reported about the concentration camps is accurate, then Germany has lost its honor."

The proceedings of a meeting of the university senate at Göttingen on 6 June 1945 is also symptomatic. The new rector of the *Georgia Augusta* proposed that "a statement about the ethical and political problem of the times" be made, "especially in view of the atrocities in the concentration camps and the related propaganda." A physical chemist of right-wing nationalist persuasion, Arnold Eucken (1884–1950), responded by cautioning "against drifting off into the war-guilt issue." The theologian Otto Weber (1902–1966) emphasized "the necessity of a basic rethinking," whereupon the meeting was adjourned with the dry statement: "The discussion is broken off."[278] At this point, a history of mentality fails for lack of suitable sources and one would have to resort to social psychology, such as in Alexander and Margarete Mitscherlich's book on the collective inability of Germans to mourn (*Die Unfähigkeit zu trauern. Grundlagen kollektiven Verhaltens*). My own expertise does not lie in this direction so I would just like to quote a sound explanation they offer for why there were so "few signs of melancholy or sorrow among the great majority of the population":[279]

> First there is a blunt numbing of the emotions in reaction to the piles of corpses in the concentration camps, the disappearance of the German armies into captivity, the news about the millions of murders of Jews, Poles, Russians, about the murder of political opponents from among one's own number. Stunned emotions exhibit the denial; the past becomes unreal in the sense of a retreat from any willing or unwilling participation in it, it sinks down into a dream.

---

[277]This ignorance is reflected in Edward Y. Hartshorne's diary entry on 25 April 1945 (see Tent (ed.) [1998] p. 29): "Certainly one of the most amazing psychological aspects of the Nazi system was the capacity people had to protect themselves from unpleasant facts, or at least unpleasant facts in too much detail. They appear to have known just enough about the terror to be intimidated by it but not enough to have been really shocked or stirred to action." (cf. ibid., pp. 141f. and on the following Brüche's diary no. III, BLM, box 7).

[278]Minutes of the University Senate (UAG), cited in Hentschel & Rammer [2001] p. 189.

[279]Mitscherlich & Mitscherlich [1967] p. 40. Kaiser [1997] p. 241 adopts this model for science and technology after 1945, whereas Kielmansegg [1989] pp. 61ff., receives it more critically.

The London *Times* reported in May 1945 that observers entering Germany with the British troops found the population in a stupefied state.[280] The industrial physicist Max Steenbeck also reported apathy at the news about the unconditional capitulation and a general "disgust with life" among people happening to be in captivity just then. Steenbeck was working as a foreman at the Siemens & Halske company until 1945:[281]

> A fatalistic dead lethargy gripped most of them: Take it as it comes. – There was as good as no talk of personal guilt or even partial responsibility for our situation, at best only about the stupidity of not having jumped ship in time.

The publisher and editor of the *Neue Physikalische Blätter* wrote in their introduction to the first postwar issue about the "shock effect of waking up in an uncannily changed world."[282] Hilde Lemcke complained in a letter to Meitner: "There is no impetus, pleasure in working or 'reason why'."[283] In the midst of ruined Hamburg Brüche noted in 1948: "Is there any sense at all in working on the reconstruction? Shouldn't one just rather hide away and try to eke out a living with a minimum of effort—as an egoist, like so very many other people? I can't see any of it anymore. One gets so tired."

The victims of Nazi persecution also joined the chorus of embittered, orientationless Germans, for different reasons: "Absolutely nothing is being done here for the 'victims of fascism,' so the feeling of emotional bitterness is added to the material worries and the great uncertainty defining the general situation. I am still trying all sorts of things to be able to leave Germany."[284] This was compounded by a deafness among Germans toward the sufferings of the persecuted. Any remnant compassion was stifled by the never-ending horror stories and pictures from the extermination camps. The *Neue Rundschau* of July 1946, at that time still being printed in Stockholm, reports:[285]

---

[280] *The Times*, London, 18 May 1945; cf. Balfour & Mair [1956] p. 56: "state of mental daze." Marshall [1980] p. 662 reports: "The overall response of the German population [...] was lethargy."

[281] See Steenbeck [1977] pp. 151, 156f. and 161; cf. also Hartshorne in Tent (ed.) [1998] p. 96: "It seems that almost all the characters that have survived these painful times in Germany are vollkommen zermürbt [completely crushed]." See also Arendt [1993] pp. 24f. about the apathy, indifference, and inability to show emotion among the German public.

[282] *PB* **2** (1946) p. 2.

[283] Still written in Tailfingen, 5 Dec. 1946, quoted in Lemmerich (ed.) [1998] p. 472.

[284] Ursula Martius to Otto Hahn, 16 Dec. 1946, AMPG, II. div., rep. 14A, no. 2726 (I am grateful to Gerhard Rammer for pointing out this letter to me).

[285] Eugen Kogen in July 1946 in *Neue Rundschau*, cf. Schlüpmann [2002] pp. 400f.

Progress today is severely blocked by an oppressive atmosphere of resent-
ment and repressed feelings. [...] A people who had seen the charcoaled
remains of their wives and children everywhere in the air-raided cities could
not be shaken by the amassed piles of naked corpses presented to them
recently from the concentration camps; and they were only too inclined to
look at the dead strangers and outcasts with a hardened heart and less compassion than for their own flesh and blood, killed in the rain of phosphorus
and hail of shell fragments. [...] Most of the Germans liberated from the
camps did their bit to extinguish the last flicker of sympathy for them. A
stoic minority of them quietly set out on their new paths—disappointed
about the type of "better world" supposedly in the making, for which
they had fought and suffered. They silently work and wait. The majority,
however, had nothing but complaints, abuse, and demands of the German
people. [...] Thus it happened that I could meet people who cold-bloodedly
thought it would have been better if all the liberated "Kazettler" had gone
under! And no reasonable person in Germany can refrain from sponta-
neously shrinking back from us when he hears the notorious sound "KZ"
[concentration camp]! What might have been the beginning of serious
rethinking has become a mental obstacle to inner renewal.

Fear of having to suffer punishment for the mass-murders committed in the
name of the German people probably led to the total rejection of any mention of
collective guilt. Anyone daring to do so, such as Pastor Martin Niemöller (1892–
1984) during a speech at the University of Erlangen at the end of January 1946, was
hooted down and taken for a compliant lacky of the occupiers.[286] Concrete fear
of punishment or personal repercussions, such as job dismissals, condemned to
failure any attempt to come to terms with collective involvement by the German
nation with the Nazi ideology. Although not applicable to each individual case,
complicity thus became the more permanent, cognitively and emotionally. It also
explains Eucken's warning against "drifting off into the war-guilt issue" quoted
above from the minutes of the university senate in June 1945.

---

[286] See *Die Neue Zeitung*, 1 Feb. 1946, S. 4: "Demonstration gegen Niemöller." Tent [1982] alludes
to the appearance of graffiti after his talk lambasting him as a "tool of the Allies." Schlüpmann
[2002] points out the surrounding circumstances of Niemöller's talk (arrival in an official vehicle
of the Allies, organ music setting the stage for his entrance into the church), which were hardly
conducive to a good reception of his message. The animosity that Alexander Mitscherlich was
subjected to after he and his coworker Fred Mielke published a documentary volume about the
trials of Nazi medical doctors was similarly motivated. Cf. the debate between F. H. Rein and A.
Mitscherlich on science and inhumanity: "Wissenschaft und Unmenschlichkeit," *GUZ*, 3rd ser., no.
14, 20 Jun. 1946, pp. 3–5 and no. 17/18, p. 6. From today's point of view, Mitscherlich's stated
purpose was entirely appropriate: "We do not intend to throw light on the guilt of individual
men—which is not our place—but to make perceptible one portion of the overall connections at
work in our times, which all nations are painfully caught up in."

No one saw this connection more clearly than Lise Meitner. At the beginning of 1947 she also tried to explain it to Max von Laue:[287]

> What seems so unsettling to me, in Germany's interest, is the circumstance that all this rejection of the abhorrent Nazistic culture and any professions of what German culture once truly used to mean, come either from people out of the concentration camps or from emigrés like Rauschning [...]. In England books are being written in which the conduct of the English and the Americans are being most scathingly criticized, without the author risking being discriminated against as un-English. In Germany it seems to be regarded as betrayal of the German cause if anyone were to say how terrible the past events had been.

Military strategy and economic reasons within the context of the Cold War soon moved the Allies to exchange the assumption of collective guilt with a sort of "collective innocence."[288] The wobbly lid being determinedly held down over the boiling concoction of ideological set pieces from the Nazi stage and newer ideas from Adenauer's Germany finally flew off in the explosive student revolt of the late 1960s.

In 1948 Samuel Goudsmit asked the editor of the *Physikalische Blätter* why no "revelatory reports about the damages that German science had incurred from the Nazi doctrine" had appeared yet besides Ramsauer's article. Brüche agreed that such articles would be very desirable, but pointed out that "a German scientist still [had] to overcome great inhibitions before he could put pen to paper on these questions." The subsequent passage illustrates the mental mix of intellectual paralysis, disenchantment, and hopelessness:[289]

> The task of overcoming the lack of understanding for the German situation seems pitifully hopeless. Let me just recall that Mr. Einstein recently denied the League of Nations his participation because it was German business and that a survivor of Auschwitz had to tell him that by doing so he was not following his own precepts. Mr. Einstein failed to see that National Socialism had been fighting not just the world but also the German intelligentsia, which is now struggling to reconnect with the international world.

---

[287] Meitner to von Laue, 13 Jan. 1947, cited in Lemmerich (ed.) [1998] p. 480.

[288] This idea comes from Schlüpmann [2002] p. 430, citing Hans Habe; cf. Balfour & Mair [1956] pp. 60f. for an evaluation of the thesis of collective guilt from the perspective of the Allies. German emigré Jews raised protest against this thesis: Engelmann [1987] pp. 161f. Frei in König et al. (eds.) [1997] pp. 220f. also criticizes it, pointing out a collective escape from responsibility and denying the supposed "right to forget" claimed by the "highly conformistic society of the Adenauer period."

[289] E. Brüche to S. Goudsmit, 20 Sep. 1948 (carbon copy), BLM, box 104, folder 261.

> Thus disappointment is beginning to spread among German scientists that
> their hopes for an era of science without Hitler have not been fulfilled.

Driven to the defensive, this German thus inadvertently replied with a scarcely disguised counter-accusation. It was in response to a carefully and politely phrased inquiry by a physicist in the US thoroughly acquainted with the European situation. Setbacks and difficulties encountered in hesitant attempts at dialog, such as Albert Einstein's (1879–1955) brusque rejection mentioned in this quote, engendered deep disappointment and frustration.[290] Other elements of the mental landscape of the time, such as a sense of isolation, travel restraints, and general insecurity, combined to form a mixture of listlessness, resignation, and indignation, making virtually impossible a meeting of minds, let alone reconciliation with exiles or representatives of the victorious powers. My allusion here to a coherent group of "German scientists" with commonly held attitudes recurs, incidentally, in many primary sources and is therefore accurately described as an "actor's category."

---

[290] His confidence in the efficacy of such international endeavors was considerably weakened by this time, based on his experiences as a member of the *International Committee on Intellectual Cooperation* of the League of Nations. See, e.g., Grundmann [2005] section 2.7.

10

SELF-JUSTIFICATION AND THE GUILT ISSUE

In 1947 Ernst Brüche asked a former colleague of his, who had been working since his emigration in the electron microscopy department of RCA Laboratories in Princeton, what the general attitude in the United States was toward Germany. The reply was that there was widespread distrust: "This distrust is based, among other things, on the fact that after the collapse almost no one wanted to admit that he was a National Socialist or had supported Hitler's government and that as good as no sense of guilt exists."[291]

The general initial response of Germans was to push aside the past and not think about it. But there were a few early attempts to give more or less explicit justification for specific action—or inaction. They were usually in response to questions posed from the *outside* (e.g., by Allied Control officers or foreign friends), such as why German scientists had put up so little resistance. Such uncomfortable questions would not be posed by members of their *own* reference group. At least by the time denazification was formalized (e.g., implementation of the Law for the Liberation from National Socialism and Militarism, in the American zone on 5 March 1946), self-justification had become a versed and almost automatic practice which, as the social historian Cornelia Rauh-Kühne noted, could only be "highly detrimental to the political ethics of postwar society." The persons affected by the law were classified under one of five categories of involvement:

I   major offender
II   incriminated activist, militant, or profiteer
III   lesser offender
IV   fellow traveler (i.e., nominal supporter)
V   exonerated individual

Contrary to normal judicial procedure, the burden of proof rested on the defendant. He had to try to refute the legally imposed presumption of guilt, which was mostly based purely on formal membership in the Nazi party or in one of

---

[291] Edward Raneberg to E. Brüche, 30 Aug. 1947, BLM, box 119, folder 308. On the sense of guilt cf. the questionnaires in Merritt & Merritt (eds.) [1970], [1980] pp. 146, 197f.

its affiliates.[292]   Another classificatory scheme suggested by the playwright Carl Zuckmayer, in exile since 1939, would have agreed much better with contemporary sentiments. It appeared in a secret report first published in 2002 outlining approximately 150 character profiles of writers, publishers, and other members of the fine arts:[293]

> 1. Active Nazis and mean-spirited followers [...], who joined the Nazis against their own convictions and better judgment [...] 2. Gullible followers who could not resist the spell of the Nazis [...] but still tried to remain decent personally. 3. Indifferent and helpless persons [...] 4. Conscious advocates of inner resistance.

Practically speaking, this character scale would also have generated an inflation of "gullible followers" and "advocates of inner resistance," of whom there were suddenly so astonishingly many after April 1945.

Within the context of a preliminary "reopening of professional contacts," a written exchange took place between a physicist at Tübingen and his Parisian colleagues at the end of 1945. The German wrote a letter, upon returning home from a visit to Paris, in an effort to respond more fully to questions that his French hosts had raised. He was evidently so confident of the cogency of his letter that he immediately published it anonymously in the daily *Schwäbische Tagblatt*. Brüche also reprinted it in the *Physikalische Blätter* under the title "Letter to France."[294] In countering the frequently voiced opinion "that German scientists could have opposed the tyranny much more resolutely and energetically," the letter writer (who incidentally revealed his identity two years later as Brüche himself)[295] pointed out "that National Socialism initially left a not intolerable impression," even outside

---

[292]See Rauh-Kühne [1995] p. 51. A facsimile of the form drawn up on the basis of the Law for the Liberation from National Socialism and Militarism, obligating all Germans over 18 years of age to fill it out, is accessible at www.lrz-muenchen.de/~Sommerfeld/gif100/04727_01.gif and _02.gif The first steps toward a more differentiated classification of the degree of political incrimination (see, e.g., Tent (ed.) [1998] p. 103 for a proposal in the British zone) or of personal guilt (ibid., p. 56) were unfortunately not followed through.

[293]See Zuckmayer [2002] p. 13. Hückel [1975] p. 163 criticized such "senseless political questions and requirements" as exclusion of former members of the Nazi cavalry corps.

[294]"Brief nach Frankreich" *PB* **2** (1946) pp. 8–11, quote on p. 10.

[295]See *PB* **4** (1948) no. 8 (submission closing date: 1 Nov. 1948), p. 360: E. Brüche: "Offener Brief an einen Franzosen" (First Lieutenant Robert), "and my 'letter to France' showed you ...." The sequel open letter revealed how much the climate had deteriorated in the course of those two years: "no careful observer will be deceived that pragmatic nationalism compliments this theoretical internationalism. A cynic would say that for the wielder of power both aren't so disagreeable either, because he can also dictate the rules on how the dispossessed may speak. It is, as a matter of fact, understandable that your government would have the desire to improve France's economic

the country. Not a single nation broke off its diplomatic relations with Nazi Germany, for instance. There then followed the familiar litany of justifications: about the "hollow desperation of the German nation" since the global economic collapse and its readiness for change at any price; the unpredictability of Hitler's actions after he had taken over government; and the rapid destruction of all democratic rights after which "any resistance from among us" would have meant "certain suicide." The whole prelude to this publication casts doubt on whether the sole purpose had been to convince French colleagues of this apology, who would rather have been repelled by the publicity. The main intended audience was domestic. A particular image of scientists fighting to the last against the Nazi party and state was being suggested to the German readership.[296]

> But there were other arguments [against resistance] as well. Many long hoped for a reorganization from within and it is sure that many men fought a long, desperate, and dangerous battle for an inner purging of the party and the state. It depresses us that even these men, to the extent that they were party members, are now being processed by the "denazification machine." Others among us Germans confidently expected that this "sickness" would self-destruct through intrigue, palace revolution, and assassination.

This rhetorical figure recurs frequently. But any overly strong influence on matters of science policy by party officials and other persons alien to science generally encountered resistance, and such resistance was locally and functionally constrained. Postwar revisions exaggerated it as opposition to the Nazi regime as a whole. "The feeling of being victimized overlay awareness of having cooperated with National Socialism and of having deserted their persecuted fellow Germans. A 'personal reality' of the Nazi past was constructed that lost touch with the true circumstances as it retreated into its own fiction."[297]

Another form of self-deception particularly popular among physicists was to think that they had succeeded in harnessing the war machine for their own purposes. In this rendition, the military was a useful bill-payer, particularly for prestigeous large-scale projects requiring expensive instruments or lengthy basic research. A transcript of Samuel Goudsmit's interview in April 1945 with Hans Kopfermann (1895–1963) and Fritz Houtermans (1903–1966), who happened also to be around, about their work for the *Uranverein* is an example. Kopfermann

---

situation, but a 'sacro egoismo' is questionable that books as credit to its own nation whatever damages the other."

[296] *PB* **2** (1946) p. 11.

[297] Szabó [2000] p. 508.

replied that he had only been peripherally involved in this pilot nuclear research program and hardly knew anything about what the others had been working on. His own research concentrated on mass spectrographs and hyperfine structure. Houtermans, by contrast, who had been ordered to work on the project, attempted repeatedly to pass information on to the Allies at the time about the latest results. As different as the two were in their political views and attitudes and in how they had acted, they agreed on one thing:[298]

> Both K[opfermann]. and H[outermans]. confirmed the general attitude of physicists towards the war effort. Independently they quoted the same statement we had heard before, namely that they "put the war in the service of science" and not the other way around. This, Houtermans said, is a direct quotation from Heisenberg.

Another example is financing by the Air Force of major tower telescopes for solar observation. Sunspot activity leads to a temporary breakdown in radio communications, so predictions of such disturbances were militarily important. Other physicists employed in the *Uranverein* truly thought they could use the Army's momentary interest in building an atomic bomb for doing basic research on nuclear physics.[299] Such cooperation for private ends was essentially collaborating with the Nazi regime and only some physicists eventually realized this. Carl Friedrich von Weizsäcker, for instance, only admitted this publicly in the late 1980s during an interview with a newspaper reporter.

Another variant of self-deception can be found in the turncoat Wilhelm Westphal's postwar correspondence with James Franck, an exile since resigning his professorship in 1933. After Westphal had written him three letters in 1947, Franck summarized his attempts at justifying himself in a letter to Lise Meitner:[300]

> Westphal wrote me from America, where he had been sent as an expert. Reading his letter one gets the impression that he had done wonders of heroism and had nothing to reproach himself for. I supposedly had been unfair to break off our old friendship. His letter was written in genuine

---

[298] Cited from the transcript: Conversation with Houtermans and Kopfermann, April 17, 1845 at Göttingen. Major R. R. Furman and Dr. S. A. Goudsmit, National Archives, Washington, RG 77 entry 22, box 167, signed by Samuel Goudsmit; Gerhard Rammer kindly pointed out this document to me. Houtermans's career was extremely unusual. See, e.g., Landrock [2003] and here p. 126.

[299] On the former (albeit not critical enough of the astronomers' version): Wolfschmidt [1992]. On the the the latter, in self-critical retrospect: von Weizsäcker [1991] and Walker [1990].

[300] J. Franck to L. Meitner, 18 Jan. 1947, Franck papers, RLUC, box 5, folder 5. See also ibid., handwritten draft dated 25 Jan. 1947, sheet 2, and Westphal's correspondence with Franck, ibid., box 10, folder 4, esp. Westphal's letters from 23 July and 22 Oct. 1947.

sincerity; he believes what he says. I just think that he is one of those lucky people who never have a clear picture of themselves. I answered him but could not remind him of a number of things he had done because of the censorship the letter would undergo and just said that I hoped to have the opportunity some day to explain my position to him thoroughly.

Instead of accusing his former friend of conscious deceit, Franck very rightly recognized that Westphal, like so many of his other colleagues who had stayed in Germany, was suffering from deep-seated self-delusion. Many documents, including almost all the requests for exonerating certificates, reflect it.

Brüche's letter to France offering an apology for the political inefficacy of German physicists blames the cliché of the anemic intellectual. This "subtle propaganda" had accordingly led to "alienation from among the professional strata" the "full extent of which still needs to be recognized today." He continued: "Many of my colleagues and I myself encounter daily the obstinate arrogance and self-secure banality with which subordinate officials and large segments of the population dismiss science as superfluous or generally dangerous."[301] This time the scapegoat was "anti-science Nazi propaganda," with the implicit admission of its strong after-effects. The moral of this tale is clear: give scientists better press coverage, power, and influence, then you will have a better world. So engrained was this ideal that the author never even noticed that he was simply exchanging one *Führer* ideal for another: the "intellectual leader." I note here in passing that the successor to the Kaiser Wilhelm Society, the renamed Max Planck Society, continues to institutionalize this *Führer* ideal, being organized around prominent personalities. The Allied occupiers raised objection to this "Harnack principle."[302]

It is also remarkable that the unconditional sense of duty pervading Max Planck's presidency of the society during the Nazi period until 1937 was continued by his successor. Otto Hahn assumed this office on 1 April 1946. His article discussing the Max Planck Society as "a research institution without state or economic tutelage" ends with a quote by Max Planck about the "pure mentality, which finds its expression in conscientious fulfillment of one's duty."[303] As if countless crimes by white-collar workers of the likes of Eichmann had not been committed in exactly this spirit.

---

[301] *PB* **2** (1946) pp. 8–11, quote on p. 11.

[302] For instance, Murray [1949a] p. 170 in the name of a delegation of the British *Association of Scientific Workers*: "The self-governance clauses in this society's by-laws are extremely undemocratic. They guarantee that the management stay in the hands of a small group of elder scientists or of those whose views they condone." Cf. also Hahn [1949] pp. 3f. for a response and AMPG, div. III, 14A, no. 6204 for a similar response to an article in a British paper in 1948.

[303] Otto Hahn: "Die Max Planck-Gesellschaft," *GUZ*, 3, no. 7/8, 19 Mar. 1948, pp. 14–15.

How did the behavior of people differ who were known abroad for their unbending attitude toward the Nazis and were regarded as representatives of a better Germany? Could not at least ethical and independent minds like Max von Laue escape the contemporary mental corset? His protest letter to the *Bulletin of the Atomic Scientists* from 1948 in response to a book review by the American physicist Philipp Morrison (born 1915) is an appropriate example. The book *Alsos* is an eye-witness account by Samuel Goudsmit, leader of the special corps responsible for ascertaining how far the German nuclear research program had come toward building an atomic bomb and for tracking down the people involved in it. Although the mission itself was very successful, the book had obvious weaknesses. Goudsmit had a tendency to paint everything either black or white. He downplayed German accomplishments in nuclear research and too eagerly pointed an accusing finger at anyone, whether or not directly involved with the Nazi machine of extermination. But was it necessary for von Laue to respond so heftily to Morrison's review, which even conceded that Goudsmit had not done full justice to those who had tried to stay uninvolved in war research? Max von Laue wrote:[304]

> One concludes [...] that it is the reviewer who explicitly puts forward the monstrous suggestion that German scientists as a body worked for Himmler and Auschwitz. How far Morrison has suffered personally through both or either is unknown to us. We do know that Goudsmit lost not only father and mother but many near relatives as well, in Auschwitz and other concentrations camps. We realize fully what unutterable pain the mere word Auschwitz must always evoke in him. But for that very reason we can recognize neither him, nor his reviewer Morrison, as capable of an unbiased judgment of the particular circumstances of the present case.

---

[304] M. von Laue [1948] p. 103 and pp. 424–425; cf. Hentschel (ed.) & Hentschel (ed. asst./trans.) [1996], doc. 117. Herbert Jehle also complained about it in a letter written in the USA to Karl-Friedrich Bonhoeffer in April 1952 (AMPG, III, rep. 23, no. 30,7): "Such reports as 'Alsos' are to blame for the distorted picture repeatedly presented to Americans. The connection between the sciences and the resistance movement [...] is virtually unknown." Sime [1996] pp. 493f. quotes from Paul Rosbaud's letter to S. Goudsmit, 13 Nov. 1949, that Fleischmann had told Rosbaud, "almost *everybody* in Germany regarded *Alsos* as the most infamous book ever written" (original emphasis).

This text turns a German who had no need to come to his own defense into a self-assigned judge. Max von Laue took it upon himself to pass judgment on who could or could not decide the special circumstances of the case at hand, that is, who could distinguish reliable testimony from prejudice. As if he alone stood on the moral high ground. Like a cornered animal he tried to parry the aggressive attack on German scientists as a whole with a frontal assault. What made him feel so personally affected? Other readers privately shared similar views. The person Morrison had singled out as the "only" honest man was one of them: Wolfgang Gentner (1906–1980). He wrote to von Laue in 1948:[305]

> Having been specifically named, I may surely convey to you my complete agreement with your reply. I am of exactly the same opinion that one cannot discount the attitude of German physicists in such simple terms as Mr. Goudsmit has evidently chosen. The verdict also seems very superficial and does not attest to sufficient knowledge of the situation for passing such severe judgment.

Morrison responded to the attacks on his review of Goudsmit's book by pointing out the inconsistency of von Laue's argumentation:[306]

> Professor von Laue, whose outspoken opposition under Hitler was a token of his wisdom and integrity, begins his piece with a most uncharacteristic reference *ad hominem*. He wonders whether I suffered personally through Himmler and Auschwitz. I did not. But I do not see that it is fair or relevant to ask. I am of the opinion that it is not Professor Goudsmit who cannot be unbiased, not he who most surely should feel an unutterable pain when the word Auschwitz is mentioned, but many a famous German physicist in Göttingen today, many a man of insight and of responsibility, who could live for a decade in the Third Reich, and never once risk his position of comfort and authority in real opposition to the men who could build that infamous place of death.

Eugene Rabinowitch (1904–1973), the editor of *Bulletin of the Atomic Scientists*, founded in 1945 as a mouthpiece for public reflection on the consequences of nuclear weapons, emphasized that no one was blaming *all* German scientists.

---

[305]W. Gentner to M. von Laue, 10 May 1948, AMPG, div. III, rep. 50, no. 688. Gentner was involved in drafting the German Physical Society's letter of apology to former members who had been instructed to cancel their memberships in 1938. See ibid., von Laue to Gentner, 3 Nov. 1952 (forwarding of von Laue's draft), and Stefan Wolff in Hoffmann & Walker (eds.) [2007].

[306]P. Morrison (followed by Rabinowitch) in *Bulletin of the Atomic Scientists* 4, no. 4 (1948) p. 104. Cf. Hentschel (ed.) & Hentschel (ed. asst./trans.) [1996] doc. 118. Unfortunately the *Bulletin*'s files (RLUC) contain virtually no other correspondence with the editor about this.

Nevertheless, considerable difference remains between his [Morrison's] qualified readiness to extend the hand of friendship to many and von Laue's feeling of solidarity with the majority of his colleagues.

Von Laue would like to see the notorious Nazis among the scientists, or their friends, punished or at least morally isolated, and, at the same time, professional recognition and human rights restored to all those who merely carried on as best they could under the tyrannical regime. Morrison, on the other hand, is inclined to include many of these "average" German physicists, chemists, or engineers under the "not unwilling armorers of the Nazis." Even more pointedly, his criticism applies to some men of great prominence in science, whose apparently unreserved collaboration with the criminals in power undoubtedly helped to subdue the qualms of many a rank-and-file German scientist.

The next passage of his editorial explains why the book and the judgments contained in it had made such waves. Individuals were not being scrutinized so much as the basic position taken by the Allies on German scientists as a group. That was why in the above quote Rabinowitch drew chemists and engineers alongside physicists—incidentally another indication that these facets of mentality were not confined to the physics community:

If things are allowed to take their course, the international scientific community will undoubtedly "long delay"—to use Morrison's expression—the welcoming of German scientists, even purged of notorious Nazis, into international organizations, and conferences. After the war of 1914–1918, several years had to elapse before the bitterness between scientists and intellectuals on both sides died down; it will take many more years after the Second World War for time to heal the much deeper wounds.

Ties between the Allies and Germans were still broken at this point in time (1948) but Rabinowitch's prediction would not become reality. The Cold War forced the building of bridges over old chasms in the West and East. But this lies beyond the temporal scope of the present study. At the close of the 1940s assessments of the situation still differed widely.

Lise Meitner gave an initial tentative assessment of von Laue's article in the *Bulletin of the Atomic Scientists* in a letter to James Franck:[307]

---

[307] 27 June 1948, Franck papers, RLUC, box 5, folder 5. The journal's files contain only one letter about von Laue's article. C. Sheldon Hart from Northfield, Minnesota, picked up where von Laue had left off on the issue of guilt: "Mr. Morrison does not seem to be aware that since the United States is preparing to use atomic and germ warfare, our own position, morally, will not differ from that of the Nazi scientists. Therefore, it behooves all of us to be charitable to our fellows in other lands; for our own turn is coming." No response by the editor has been preserved.

it seemed to me very naïve and certainly not suited to the purpose Laue had written it for. What do you think? Perhaps you think I should let all these things be, but it is precisely because I feel so strongly attached to Germany that I cannot stop thinking how unfortunate the attitude of most academics is for a better future. Maybe I am wrong?

On more than one occasion Meitner and other exiles were disappointed on the personal level when former colleagues attempted to resume contacts as if nothing had happened in the interim. Walter Grotrian (1890–1954), for instance, had been serving as an officer in the Air Force during the Nazi period in Norway but did not give the least indication of having given it the slightest thought, let alone felt any regret or remorse:[308]

> As regards Grotrian, the Hahns earnestly contended that the role he was able to play and not least the splendid uniform he got had determined his change in attitude. I told the Hahns many times that I did not believe this explanation but they insisted on it. I still don't believe it. The Hahns are far too uncontemplative to be experts on human nature. But I do not know what really led Grotrian to change his mentality. Shortly before I left Germany he visited me and was simply in despair about the Nazis. He isn't particularly clever or capable of deep thought; but I find it lamentable that he would write you without touching on the past. It reveals a lack of honesty that I would not have expected of him, but perhaps it is only a lack of courage of conviction.

When Grotrian also contacted Meitner directly about a year later, she frankly told him that she had heard about his appearances in Norway and could not understand it. Grotrian tried to justify himself in his reply:[309]

> Much as I deplored the outbreak of the war, once it was a fact, I could not resolve to refuse my services to my fatherland. In that, I saw no support of Nazi goals, but fulfillment of a duty that stood above political antipathies. [...] I never believed, even during the period of the greatest military successes, that we would win, but I hoped that we would not be

---

[308] L. Meitner to J. Franck, 16 Jan. 1946, RLUC, box 5, folder 5. Grotrian's mission in Norway, involving among other things ionospheric measurements at the local northern lights observatory for the purposes of the *Luftwaffe*, is discussed by Hoffmann in Dörries (ed.) [2005]. Grotrian's later letter of justification to Meitner 1947 is also quoted below.

[309] W. Grotrian to L. Meitner, 29 Dec. 1947, Meitner papers, Cambridge, box 5/6, folder Grotrian, quoted from Hoffmann in Dörries (ed.) [2005] pp. 42f. Grotrian seems to have suspected that Meitner would not be convinced by his arguments and added: "It is clear to me with your completely different kind of attitude, you are unlikely to understand my way of acting." Even this perceptiveness is missing in many other similar letters of the time.

totally defeated and contributed in my direction, setting aside grave doubts. There was, after all, only a choice between various evils, and it remained an open question which of them was greater. I took total capitulation of Germany as the greater evil, and now, after it has happened, I find my fear confirmed. Germany may now be freed of the Nazis, but it is, outwardly and inwardly, physically and mentally, a pile of ruins, under which lies all that was dear and valuable to us. I know that this destiny is our own fault and that we must bear it. But even now, I still cannot feel guilty for having exerted myself to stave off this fate.

Grotrian had placed himself at the service of the Nazi power-mongers and military machinery—not without some hesitation. But in 1947 he still could not feel the least guilt. As he saw it, he was working throughout not for victory but against total defeat; not for the Nazis but for Germany (adding: "at the time I had not fully grasped that the men at the top were actually criminal types"). Contrasted against such failed attempts at self-justification are the few acknowledgments of involvement by those who had fled Germany in time to be able to keep far away from the gruesome events. This distance may well have been an important aid in attaining the level of reflection in James Franck's letter to Max Born at the end of February 1953 from Uppsala. The occasion was the prospective conferral of an honorary citizenship by the City of Göttingen, which he interpreted as[310]

> that the Council of the City of Göttingen had the intention of honoring through us the millions who fell victim to the racist insanity of National Socialism. [...] Otherwise I fully agree with you. As far as my belief in communal guilt goes, I cannot exclude myself or even you—although we saw much more clearly. Under the Kaiser we always nicely kept our traps shut. I considered his speeches silly and stupid but ought to have known what impression they would make abroad and that they made a war possible.

This harmless form of self-criticism was evidently not possible—publicly, at least—even for German physicists willing to acknowledge culpability in some way or other. But it is hardly evident in their private correspondence either. There are a few documentary indications that the guilt issue was discussed much more openly in conversation, for instance, among Göttingen physicists after 1945. The

---

[310]Born papers, SBPK, quoted in Lemmerich [2002], pp. 167f. (Likewise to Gaffron: "as long as German nationalism raged in the Kaiser's speeches, we never took it seriously enough to see that it contributed hugely toward catapulting the whole world into misery.") The unpublished script of the speech Franck gave on the occasion of the award as honorary citizen of Göttingen is among the Franck papers, RLUC, box 3, folder 8, as an enclosure to Franck's letter to Chief Mayor Foge dated 21 Feb. 1953.

following passage comes from a confidential letter by Fritz Houtermans. As a member of the German Communist Party, Houtermans had to flee Germany when the Nazis took over government. In 1935 he left England to work in the Kharkov Institute of Nuclear Physics in northern Ukraine. The pact between Hitler and Stalin in 1939 led to his deportation back into the hands of the Gestapo. Houtermans barely escaped death in their custody when Max von Laue managed to have him assigned to work on war-related research as a member of Manfred von Ardenne's team. Houtermans wrote to Michael Polányi:[311]

> nobody can be served by hiding the truth. On the other hand, however, I know from long experience as a "fellow traveler" and from the reaction of many of my friends how badly they take it when things are discussed in public that they would certainly be willing to talk about in private conversation and are open to being convinced about but immediately freeze in resistance as soon as it involves public debate. They then immediately feel like "standard-bearers of the faith" and consider it their duty to support even things that they personally find completely questionable, especially when their counterpart is stamped as an "anti[-German]." Because it is important for me, and surely also for you, that these circles in particular be persuaded, I would very much like to make as few obligatory public appearances as ever possible and to have as many opportunities as possible to have private conversations. [...] I don't know whether you understand how much "mental capital" our friends have invested in their convictions and we want to make it as easy as possible for them to lose or weaken it.

Some Germans were aware that their passive resistance or emotional retreat during the Nazi regime (their "inner emigration") did not *eo ipso* absolve them from their share of responsibility for what had happened. Nevertheless—as Balfour perceptively analyzed[312]—they were stuck in a

> conflict between remorse and patriotism which often bottled up their feelings in an unconstructive way. They were loath to admit their mistake or ask for forgiveness because that might be construed as surrendering the German case. They were very ready with arguments to justify their attitude, but arguments did not help since what was lacking was moral courage rather than intellectual conviction. Indeed the Allies made a mistake with such people in harping on the theme of guilt and trying to insist on its admission. The more they insisted, the greater were the psychological barriers created.

---

[311] Letter dated 1 Oct. 1948, Polányi papers, RLUC, box 5, folder 5.

[312] Balfour & Mair [1956] p. 60. See also pp. 14–64 there for a generally fitting description of the basic attitudes and expectations held by members of all four Allied occupying countries as well as by the Germans themselves.

When in doubt, the decision on whether or not to make any public confession fell *contra*. After seeing a letter Niels Bohr (1885–1962) had written to Otto Hahn in 1946 that suggested he considered it appropriate for German scientists to apologize publicly for the treatment of scientists in countries occupied by Nazi Germany, Max von Laue responded by writing this Danish mentor of numerous quantum physicists:[313]

> I hardly believe that the Germans coming into consideration would find themselves ready to do so. In any event, I am against it. Such self-evidences are not said so specifically, least of all in a formal declaration. If our colleagues abroad would like to hear such declarations, documenting a distancing from the spirit of the Third Reich, they only need to take a look at the speeches that the presidents at German universities delivered at the inauguration of the new semester or before the newly matriculated students a few months ago, e.g., the pertinent speech by the Göttingen rector, Professor Rein. They are appearing in print—albeit perhaps not all of them—and thus hopefully also extend beyond the Reich's borders in this form. Such declarations have a specific, clear reason. A reason is missing for such a declaration you are seeking and this would easily give rise to all sorts of assumptions that one would better not have arise.

> Hahn, who has read this letter, sends his best regards.

Evidently, von Laue was afraid that such public statements could be misused by the Allies as an admission of guilt with the consequence of further punishment for German science.[314] Or else it might be interpreted psychologically as an admission of a bad conscience. The unpleasant memory of having been complicit in the plundering of the occupied territories—whether in its most harmless form of delivering lectures in a propaganda institute, if not by outright active participation in the confiscations of foreign research installations and apparatus—was thereby brushed aside.

---

[313]M. von Laue to N. Bohr, 30 July 1946, AMPG, Max von Laue papers, II. div., rep. 50, no. 303. I thank Gerhard Rammer for pointing out this letter to me. The talk by the physiologist F. H. Rein mentioned in this passage portrayed Germany as a victim of the military and would not have satisfied Bohr as an adequate confession of guilt or penitence.

[314]In this vein see also Tent (ed.) [1998] p. 14: "if we say that Nazis are criminals, then that means that Germans are criminals, and for us Germans to say that in front of foreigners would amount to a national dishonor (eine nationale Schmach)."

11

SELF-PITY, SENTIMENTALITY, AND SELFISHNESS

These three aspects are what Michael Balfour's study on the four-power control in Germany identified as general characteristics of German mentality after 1945. Hannah Arendt's synopsis after her visit from exile in 1950 is similar:[315] "The reality of destruction surrounding every German dissolves in a pensive but scarcely rooted self-pity, which rapidly evaporates again when ugly low buildings are erected on a few wide streets [...] to begin to cover up the dreary landscape and offer an abundance of provincial elegance in super-modern display windows." Edward Hartshorne's diaries also reflect these characteristics—for instance in his rejection of the idea of hiring Germans for evaluating filled-out questionnaires: "They were bubbling over with self-content. I must say I am against employing Germans at all. They are only seeking to improve their own lot and still despise the rest of the world." Officer Hartshorne was in charge of reopening universities within the American occupied zone.[316] Max Steenbeck reported about his experiences in Soviet captivity right after the unconditional surrender in May 1945:[317]

> For most of us the world had fallen apart or—for some, worse still— our personal bases for earning a living and therefore our meaning in life. Everyone came from positions of authority. Here no one wanted to listen, everyone was filled with his own, different thoughts; there was no common fate to be overcome. Everyone would have liked to talk about his own, if only he could find someone who was interested.– At first, people talked about what they had experienced before, often somewhat boastfully, often omitting some things for good reasons. But once you did start listening to someone and set demands on what was being reported, you very soon encountered the other person's worries about his future—as if you didn't have enough worries of your own. Not being able to listen anymore was the true crux.

---

[315]See Balfour [1959] p. 99, Marshall [1980] pp. 660f., Broch [1986] pp. 80f., Arendt [1993] p. 28.
[316]Hartshorne in Tent (ed.) [1998] pp. 37f. (30 April 1945); cf. ibid., p. 83: "The professors and the Rector keep whining at me about their wonderful library building [...] being occupied." See also p. 119, where Hartshorne's French counterpart M. Sauzin is quoted as saying to German professors: "You shouldn't weep in your ruins, you should work!" On the British zone, see Marshall [1980] p. 666.
[317]Steenbeck [1977] pp. 157f.

The element of self-pity is also prominent in a notice reprinted in the *Physikalische Blätter* from the Swiss paper *Basler Nachrichten*. The article quotes from a conversation Otto Hahn had with a Swiss visitor and is originally dated 10 Nov. 1946, shortly before Hahn's departure for Sweden to receive his Nobel prize:[318]

> You see, I had hoped for years for the time when we would be rid of the heavy mental burden of National Socialism, and how much I looked forward to being able to work freely and without hinderance. But now I am sitting here, a head without a body; I am not allowed to return to my institute because it lies in the French zone, and I have little idea about the other institutes, and here come new people every day wanting a job or a political exonerating certificate or whatever else. I simply cannot help these people. Formerly, I really used to be a cheerful person and was actually never pessimistic, but if people just come with demands and one can hardly move for all the restrictions, I simply cannot go on. And imagine, ludicrous though it may sound, at the moment I don't even have a sound pair of shoes to put on. So, what use is it to me if the Nobel prize is waiting for me in Sweden, which I am not allowed to pick up because I don't get a travel permit and meanwhile, I submit one application after the next for months on end in vain for a pair of shoe soles. If they would at least send me a pair of shoe soles against the Nobel prize account, then I wouldn't have to walk around with wet feet all the time.

Hahn's circumstances had been rather privileged, starting at least with his internment at the end of the war in the English manor house Farm Hall, "at Her Majesty's pleasure."[319] At the time the article was written he had just been appointed president of the Kaiser Wilhelm/Max Planck Society and was living in

---

[318] O. Hahn, *PB* **2** (1946) issue no. 8, p. 240. Cf. the carbon copy of Brüche's letter to the editors of the *Basler Nachrichten*, 30 April 1947 (AMPG, div. III, 14A, no. 2302). According to it, Hahn (and Planck) were embarrassed that it appear "in print in Germany that he was doing badly, because with a very few exceptions all his colleagues here were not doing any better." Sime [1996] p. 335 apparently also gained the impression that exaggerated self-pity characterized Hahn's correspondence from this period: "his letters to Lise were a litany of hardships: shortages of food, coffee, cocoa, cigars; problems with British and American officials; demeaning travel restrictions; apartments requisitioned by occupation forces; the deaths of Hans Geiger, Otto von Baeyer, and others hastened by poor food and difficult living conditions."

[319] An old British law permitted the head of state to intern anyone without charge or court trial. The conditions during this confinement in a former manor were good and their care was even exceptional, considering the circumstances. For further background on the interment, see the introduction to Frank (ed.) [1993]. Lise Meitner's postcard to Ladenburg at the end of December 1946 after seeing Hahn again for the first time in Stockholm documents that his physical condition had not suffered by it particularly: "But Otto & E[dith] looked astonishingly well & young." (AMPG, Ladenburg-Sammlung, VA, rep. 7).

Göttingen, a town that was virtually unscathed by the war. If something negative had to come out, why did he not talk about the much worse lot of his less prominent colleagues? Hahn preferred to complain about his leaky shoes. When the notice was reprinted in the *Physikalische Blätter*, Hahn wrote the editor Brüche to deny ever having said anything remotely similar "to a journalist." The way he phrased this denial is so awkward, though, that I suspect Hahn had said something of the sort to a *different* Swiss visitor, who had anonymously passed the story on to the *Basler Nachrichten*, whence emerged the fictitious Swiss journalist. Even if "this whole matter" had indeed been "pure fabrication," as Hahn alleges,[320] it is sound fiction and reflects the mentality of the period—our object here. The following passage from the physical chemist Hartmut Paul Kallmann's letter to an emigré is in the same vein. Kallmann had managed to survive in Germany despite his "mixed marriage" but had lost both his parents in a concentration camp:[321]

> It still depresses me when I see that in this country recognition of what is of real importance in life has still not penetrated even after these terrible years. The tough momentary situation is deplored much more than the evil of the past 10 years. It ought to be the other way round. The masses still don't know what a salvation the destruction of the Nazis was to the whole world and to Germany as well.

We gather from a report by the nuclear physicist Hans Jensen (1907–1973) to the Minister of Culture of Lower Saxony about his trip to Copenhagen and Oslo in 1948 that the Danes and Norwegians he had spoken to repeatedly complained about the one-sided and exaggerated self-pity they noticed among the Germans they knew:[322]

> Many expressed their astonishment that letters from Germany and conversations with German guests always place the entire focus on the special desperate conditions in Germany; and very few noticeable attempts are made to regard the problems at least within a European context and as consequences of the past 15 years; and that there is rarely even any perceptible shame that the Nazis had wreaked so much havoc and misfortune in Germany's name. As much as I tried to explain such a closed-minded view

---

[320] O. Hahn to E. Brüche, 18 March 1947, BLM, box 119, folder 308. Contemporaries took the article quite seriously because it referred to the widespread problem of leather and rubber shortages: "The result of this article is that I receive offers for second-hand shoes from various quarters, other people commiserate with me about my desolate situation."

[321] H. Kallmann to M. Polányi, 22 May 1946, Polányi papers, RLUC, box 5, folder 2.

[322] Hans Jensen's report to the Minister of Culture of Lower Saxony, end of 1948 (carbon copy in the personnel file Jensen, Universitätsarchiv Heidelberg), quoted from Schlüpmann [2002] p. 440.

by the truly dire need these last few years inside Germany, I do have to say that this astonishment, which does appear to me somewhat justified, did make me feel quite awkward.

Lise Meitner's letters confirm this impression of saturated self-pity in Germany. In 1947, for instance, she wrote James Franck about her recent trip to England:[323]

> I recently had a chance to speak with many English persons who had been in Germany to attend German physics conferences—hence certainly good-willed about helping Germany retrieve decent living conditions. They all said to me that the Germans were full of "self-pity," saw nothing but their own predicament, always told the English first of all that they had not been Nazis. One of the English physicists [...] told me that Hahn and Laue and many others, he also mentioned Kopfermann, were absolutely not able to think beyond their own professional group. He only excepted Heisenberg and Weizsäcker and thought that even though they perhaps had been fascinated by the successes of Nazism, they did now have much broader points of view. [...] Heisenberg is certainly much more contemplative than Hahn and probably also smarter than Laue, he is also much younger—but is he sincere?

Margrethe Bohr (1890–1984) also met many significant Germans visiting her husband (before and after 1945) including Heisenberg and von Weizsäcker. She wrote to Meitner in 1948: "It is a difficult problem with the Germans, very difficult to come to a deep understanding with them, as they are always first of all sorry for themselves."[324] The continuation of the open letter by an anonymous scientist, previously cited, is a good example of how easily self-pity and self-justification could slip into melodrama:[325]

> All this was reason enough for us not to grab hold of the flywheel with our bare hands. The mark of the true martyr, who voluntarily and consciously takes suffering upon himself, is a firm bond with a belief system. Scientists are almost never of this very rare type of person. Researchers in Germany as elsewhere see as their goal a life of activity and not mortal self-sacrifice.

At this point the author apparently noticed how dangerously close he was to slipping into self-pity and tried to rationalize it away:

---

[323] L. Meitner to J. Franck, Franck papers, RLUC, box 5, folder 5; cf. also Schüring [2006] pp. 302f.

[324] M. Bohr to L. Meitner, 10 June 1948, quoted from Sime [1996] p. 358. Meyenn (ed.) [1993] pp. 545f. quotes Wolfgang Pauli's letter to Markus Fierz from 17 July 1948: "By contrast von Weizsäcker's mental attitude seems to me to be 'German' in a very negative sense. [...] It is interesting that a year ago, Mr. von Laue described Weizsäcker to me in a private conversation as 'Heisenberg's evil spirit' (also in the political sense)."

[325] "Ein Brief nach Frankreich," *PB* **2** (1946), pp. 8–11, quote on p. 10.

With these explanations I must trust that you [the reader] will not suspect that I am making an appeal to sentimentality. On one hand, this fine word from French Romanticism, which means "understanding everything" as well as "forgiving everything," is not appropriate for our world—human understanding is important, not for the sake of forgiving but for doing better. On the other hand, such a suspicion could create very painful circumstantial conditions for us ...

12

"PROPAGANDA-FREE DAY-TO-DAY" AND POLITICAL APATHY

The 'demise of ideology' that many had confidently forecast was an essential building block of postwar discourse.[326] It also cast its shadow over the Cold War policy of the West. As we have seen, physicists throughout Germany placed politicized Aryan physics rhetorically at odds with apolitical science as a (self-) justificatory strategy. In the West, anti-Communism was soon incorporated into this discourse by stigmatizing Communism and Nazism as merely two varieties of a totalitarian regime. Pascual Jordan reasoned that ideology was generally based on the mistaken belief that society could be organized according to abstract principles. This metaphysical and a-priori approach, he argued, misrepresented reality, which was why the "demise of ideology" was called for.[327] Jordan (unlike most other physicists) personally drew the conclusion that he must actively support Adenauer's conservative party, the Christian Democratic Union. But the Social Democrats also shared this anti-Communist attitude and felt the need to blacken the image of ideology per se. Many physicists saw the rigor of the scientific method as the ideal for visionary planning for the future, guided by facts not by totalitarian ideology.[328]

The ubiquity in the print media of *Lingua Tertii Imperii* (LTI), as Victor Klemperer termed the idiom of the Third Reich, created a yearning for language thoroughly purified of such phraseology, an "ideology-free discourse" in today's new phraseology. "We must part with untruths and slogans," one student demanded in the first issue of *Göttinger Universitätszeitung*. The "respectable measure of clarity and honesty under the watchful eye of the academic public" ought to "safeguard the newspaper from the dangers of the free press."[329]

---

[326] See, for instance, the analysis of speeches by university presidents right after the war in Wolgast [2001] part C, esp. pp. 325–328.

[327] "Das Ende der Ideologie": thus the title of chap. 6 of Jordan [1956] pp. 120–156, esp. pp. 141f. Cf. also Merritt & Merritt (eds.) [1970] pp. 43ff., 83f., Beyler [1994] pp. 523ff. or Sontheimer [1991] p. 143 on the anti-ideological tenor of the young Federal Republic of Germany.

[328] See, e.g., Beyler [2003] p. 228 (not just appertaining to Pascual Jordan). Schüring [2006] pp. 269–273 exposes this retreat to professionalism and strict factuality as an apologetic strategy, obscuring the circumstance that these very attributes had contributed significantly to the efficiency of the Nazi system.

[329] Wolfgang Zippel, law student, "Zum Geleit," *GUZ*, 1st ser., no. 1, 11 Dec. 1945, p. 1. The dangers evidently refer to a relapse into the old jargon.

Not that it was easy to do so that quickly. It is difficult to abandon an existing linguistic world overnight. The quote in the title to this chapter originates from an article addressed to students by State Minister A. D. Adolf Grimme (1889–1963), with the heading: "Study instead of demonstrating." It is not hard for us today to spy residual traces of LTI, which Grimme was so intent on leaving by the wayside. The closing paragraph under the heading "Heroism 1946" reads:[330]

> During the war students proved that they have no qualms about making sacrifices; indeed, many were those who stood up for their convictions in these past dozen years, for which they, these martyrs for a better Germany, mounted the scaffold. Today, daily routine must bear witness to their yet vibrant spirit. Heroism must now be our duty as survivors, in propaganda-free, slogan-free day-to-day. [...] to adjust now to the continuous deployment of life, to such permanent deployment for your livelihoods.

Only an orator would choose such vocabulary and syntax. The rhetorical momentum focuses on glorious words like heroism. ("...tritt jetzt die Forderung heran – des Heldentums.") Compare this against the odd syntax of Hitler's own speeches: "A German nation – we must again become." ("Wir müssen wieder werden – ein deutsches Volk.") The overtones of the Nazi propaganda machine still reecho, not just in the content but in the overall style. The omnipresent *Volksempfänger* radio broadcasts and other mass media had showered such rhetoric daily on the German public. But many were fed up with it by the end of the war.

Until 1933 the word "propaganda" still held a positive connotation and was even selected as part of the official name of a new ministry. One variant of the postwar distaste for propaganda and phraseology expressed itself in a preference for a purposefully austere choice of words and simple sentence structure. This new factuality (*neue Sachlichkeit*) characterizes the styles of Friedrich Hund (1896–1997) or Hans Kopfermann, for instance. There were, however, also smooth transitions into insensitive and euphemistic down-playing of past insufficiencies. The president of the Göttingen Academy of Sciences after 1945 took this course in his letter to Jews who had been pressured to withdraw from among the academy's illustrious ranks in 1933. He was not more specific than to allude to "deplorable circumstances." In response to protests against this evasive formulation he offered the following justification:[331]

---

[330] Adolf Grimme: "Lernen statt demonstrieren," *Die Neue Zeitung*, 2nd ser., no. 11, 8 Feb. 1946, suppl. on culture and the arts. Cf. Klemperer [1957/75] p. 11 on the "overuse of heroism, inappropriately applying a skewed interpretation of this concept."

[331] Smend to Franck, 1947, Franck papers, RLUC, box 10, folder 5. This incident is described in greater detail in Szabó [2000] pp. 511f. and Lemmerich [2002] p. 158. See here p. 154.

> Confronted with the unspeakable things that have happened elsewhere, we
> did not want to emphasize this matter unduly. Added to this, however, is
> a profound abhorrence among us for fancy words, after the abominable
> inflated verbosity during the Third Reich. There was furthermore a distaste
> for strong words about the Third Reich, which nowadays are cheaply made
> and with which the fellow travelers of the Third Reich are [...] trying to
> shout each other down as fellow travelers of the present day, now that it
> has become unrisky and advantageous to give the dead monster a few more
> kicks after the fact.

In extreme cases this aversion led to a total refusal to read the paper anymore
or to inform oneself politically. The scholar of English language and literature
Herbert Schöffler (1888–1996) pointed out in a speech before Göttingen students
on 28 October 1945: "At times like these the political trivialities are of reduced
significance; after the experiences of the years we had to endure, many are altogether weary of politics."[332] He wasn't just referring to the attitude of students.
The obligatory politicization of society during National Socialism had triggered a
"skepticism of politics from below" in the population at large which persisted into
the postwar period.[333] A reporter for the *New York Times*, who traveled through
Germany in 1947 in search of signs of a change of mood or mentality, summed
up his observations as a mixture between nationalism and political immaturity:
"There is a serious and growing Nazi undercurrent, for the moment cloaking itself
in 'nationalistic' terms. Most people, however, are political vacuums and tend to
believe the last person who talks."[334]

When a delegation of the *British Association of Scientific Workers* criticized the
"reactionary mentality at German universities," Otto Hahn attempted to defend
his fellow countrymen: "Students want to work and only want to work. [...] And
there is less danger of a revival of Nazism or a rejection of any idealistic Socialism
than of a certain cynicism against any state authority." The response by the head
of this delegation, R. C. Murray, was:[335]

---

[332] Quoted from the facsimile in Krönig & Müller [1990] p. 342. Schöffler's lecture "On the
situation" ("Zur Lage") in the large physics auditorium on 18/19th October and its repetition on
the 28th is discussed in Becker et al. (eds.) [1998] pp. 416ff.

[333] On this point see, e.g., Rammer [2004] pp. 192ff. and the references there about the student
association. A poll among 1,750 Germans between the ages of 15 and 25 in the American zone in
July 1950 found that roughly 80% categorically rejected exercising any political office; less than 50%
of these young West Germans were of the opinion that Germany could govern itself democratically:
see Merritt & Merritt (eds.) [1980] pp. 94f. and Söllner (ed.) [1986] pp. 189–191, 197.

[334] Stone [1947a] p. 5 on the "mood of Germany."

[335] See Hahn [1949] p. 3 as well as Murray [1949b].

> I would like to agree fully with Professor Hahn that the basic characteristic
> of German students and likewise of German scientists is political apathy.
> An ideological vacuum exists. Nazism is completely discredited but there
> is nothing of the same persuasive force to replace it yet. It may be under-
> standable but it is nevertheless unfortunate. If this vacuum persists for too
> long, it could prepare the ground for a new form of fascism.

The reactions of F. H. Rein and Max von Laue to this article reveal how thoroughly
they misunderstood it. Von Laue countered Murray's charge of "political apathy"
with:[336]

> Of course, this is just what Murray does not like. He wants more "public
> relations." We would like to point out, on the contrary, that "public rela-
> tions," albeit of an undesirable sort, were precisely what Hitler forced upon
> German scientists with varying degrees of success. One should therefore
> let caution reign in this regard and exercise restraint.

The editor of the *Physikalische Blätter* attributed the lack of motivation his colleagues
had for writing about political affairs or the past to[337]

> great inhibitions for a German scientist to overcome before he can take
> up his pen on these questions. He does not feel tempted to write about
> the inadequacies of a system that had proven its impracticality and repre-
> hensibleness so thoroughly and the less favorable sides of which cannot be
> mentioned without political consequences. The task of surmounting the
> lack of comprehension about the German situation seems pitifully hopeless.

The theoretical physicist Richard Becker (1887–1955) also thought that only extraordinary events of "elementary impact," such as the development of the atomic
bomb, could lure physicists out of a "zone of silence," in which, according to
him, they evidently normally seemed to stay: "We have heard this admonition to
consider the religious foundations of our culture often enough out of the mouths
of priests and philosophers. Nowadays their voices are joined by sober scientists,
who in view of the consequences of their research see themselves compelled to
leave that zone of silence about things "of which they know nothing." They step
out of the auditorium into the public in order to shake people into awareness and
educate them about a humane use of their power."[338]

---

[336] M. von Laue [1949a] p. 12.

[337] Ernst Brüche to Samuel A. Goudsmit, 20 Sep. 1948, Goudsmit papers, Washington, D.C., or
carbon copy in BLM, box 104, folder 261. On Goudsmit's decision to remain a member of the
German Physical Society, see his correspondence with Walther Gerlach, Feb., June and Nov. 1936
and with Walter Schottky, 24 June and 14 July 1936, DPGA.

[338] Richard Becker: "Gefahren der Naturforschung," *GUZ* **2**, no. 24, 21 Nov. 1947, pp. 1–2.

Becker also received cover from a physics student, who wrote an article in a later issue of the *Göttinger Universitätszeitung* about why scientists did not concern themselves about current-day issues. He also found it strange "to see so few scientists express their opinions in the *GUZ* about issues treated in it. Scientists are often accused of caring too little about burning current issues, above all political ones." This candidate of physics did not intend to discount this but tried to explain the attitude of scientists. He limited his scope to physicists, as "typical representatives of modern science" and painted the following mental profile for this paradigm:[339]

> The mentality of a physicist differs very much from that of a scholar in the humanities, a politician, a businessman or even a technician. By comparison it seems to me that he probably more closely resembles an artist. Add to that a shot of philosophy, rubbed off from the humanities. But that, I would think, is where the comparability ends.

After a column about profound and elementary curiosity being the main motivation behind physical research, the article takes a more apologetic turn:[340]

> Pardon the scientist for his curiosity, his obsessiveness, his "playing," and that he forgets the normal routine in the process. [...] His research, the issues that move him, have virtually no relation to public problems. [...] Of course he does remain bound to the events of the day but it is rarely possible for him to grasp more than their broad outlines. And this alone is not enough to join in the debate. So his restraint is not just attributable to a lack of interest but, to a good part, also to purely practical reasons.

Far be it from me to ridicule this quote, naïvely comical though it is. Some of the observations in it may well be right. What is clear, though, is that this text—the only explicit analysis of the mentality of physicists that I am aware of from this period—had an entirely different, apologetic function. Political aloofness was a traditional part of a scientist's self-image from as early as the 19th century. Its residual influence was strong even after 1933. Max von Laue's famous letter

---

[339] Rolf Hagedorn: "Verhinderte Naturwissenschaftler. Warum kümmern sie sich nicht um Gegenwartsfragen?," *GUZ* **3**, no. 20, 24 Sep. 1947, p. 13.

[340] Ibid. The tenor is quite similar in *GUZ* **4** (1949) p. 9: "The torturous conflicts of conscience about whether we should devote our time and effort to our true profession, scientific research, which does not even let itself be apportioned within regular office hours, or whether we should throw ourselves into the stressful struggle for political order, is familiar to quite many of us. The fact that the majority tends to go where we find greater practical success (efficiency) and therefore greater satisfaction is absolutely legitimate. [...] even writing down these lines seems problematic to me."

to Einstein is a good example. It was written after Einstein had tendered his resignation to the Prussian Academy of Sciences, which some of its members had been demanding because of critical statements he had made about Germany while abroad. Thus Einstein managed to forestall being officially thrown out.[341] "I surely don't need to tell you," von Laue wrote,

> how very sad I am about the events here; the worst is the complete impotence to do anything against it. Whatever causes a stir only makes the situation worse. [...] But why did you have to come forward *politically* as well! I don't intend to reproach you at all for your statements, far from it. I just think that scholars should be reserved. Political contests demand different methods and different interests than scientific research. Scholars, as a rule, only get caught up in the wheelworks.

After 1945 political aloofness received an additional positive connotation: being apolitical signaled that one was also "politically unimplicated" (in the sense of denazification)—whether or not it agreed with the facts. Spontaneous avowals of this stereotype were very rarely challenged. One exception is the following wry commentary after an editor with an obvious political record had been acquitted in 1949 on the strength of precisely this argument:[342]

> It ought to be made general knowledge in adjudications that the assertion that someone had not concerned himself with politics would have an adverse effect on his punishment, not a positive one on his sentence.

.

---

[341] Max von Laue to Albert Einstein, 15 May 1933, Collected Papers of Albert Einstein, no. 16 087. For the background on Einstein's cancellation of membership in the academy, see Kirsten & Treder (eds.) [1979] chap. VI. Cf. also Jansen [1992] who exemplifies by Heidelberg professors the political conceptions of German academics and their definition of their role in society. Harwood [1993], [2000] discusses the gradual evolution of a new politically active researcher type.

[342] Anonymous [1949], quoted after Fritzsch [1972] p. 29.

13

NEW AWARENESS OF A SCIENTIST'S RESPONSIBILITY

Political disenchantment also welled over into the *Physikalische Blätter* but its force was partly broken against Ernst Brüche's unquenchable optimism. His editorial quoted below, entitled "Physicists at the crossroads," starts with a pithy review of the apolitical attitude of scientists into which he incorporated sociobiological and historical comparisons characteristic of the mentality of the day:[343]

> Scientists are individualists and their verdicts about politics are rarely kind. In Germany today, just as 14 years ago, this relationship is not improved by talk about their having historically important obligations with repercussions for the future. With these arguments come calls for a politicization of universities, which now as before do not fail to include a menacing undercurrent of thorough dissatisfaction with past performance at universities. [...] The special distaste for politics in Germany stems perhaps from the position of its intelligentsia who had always been safely encased within the state civil service, where they feel much more strongly beholden to authority than to their own judgments. The strong tendency toward specialization which emerges within this sociological framework (similar to a termite society) is often valuable and sometimes indispensable; but if it is carried too far, then whole professional classes lose their ability to judge and assess their own interests against those of others. This may explain why state aggression encounters only very little active opposition among the intelligentsia in Germany in particular, especially when it leaves the cells of the scientific beehive alone.

This notice ends, however, with an about face: Brüche pleads for involvement in one of the many newly founded organizations fostering the internationalization of science and a greater awareness among scientists of their responsibilities:

> The lessons of the past few years are so obvious today that they are even reluctantly conceded by those who a short while ago let total encapsulation be their guiding motto. Acknowledgment of a scientist's share of responsibility in the common weal as well as in his own interest necessarily had to arise earlier in intellectually freer countries abroad and produce politically active scholars.

---

[343] E. Brüche: "Die Physiker am Scheideweg," *PB* **3** (1947) issue no. 6, pp. 208–207, closing date 5 Aug. 1947. Beyler [1996] discusses the objection to extreme specialization, so typical of the time.

Then follows a discussion of the organizational consequences for the German Physical Society:[344]

> The administrative session—a novelty in the history of the Physical Society— has issued carefully considered formulations that point out that the society must fulfill not only scientific duties but also the obligation of keeping alive a sense of common responsibility in the forming of people's lives and of advocating the freedom, truthfulness, and dignity of science. [... The] unanimity on resolutions passed at the administrative session in Stuttgart concerning these issues permits the supposition that the majority of German physicists are no longer standing at the crossroads but have already set off down the right path into the future. Just as at the founding of the Physical Society 100 years ago, the younger members will hurry ahead and the circumspect will follow.

The initiative for this change in the by-laws of the German Physical Society had come from Erich Regener. In 1947 he had introduced into the by-laws for the regional physical society of Württemberg and Baden a new article entitled "The common responsibility of scientific workers."[345] Article 2 of this by-law reads:

> From the fact that the knowledge reaped from physics exerts a growing influence on the mental attitudes of people, that furthermore practical physical findings are having an ever stronger effect on all areas of human activity, the Physical Society accepts the obligation to keep alive a sense of shared responsibility among workers in science in the shaping of people's lives. It will always support the freedom, truthfulness, and dignity of science.

Obtuse historical comparisons between post-1945 and post-1933 and quaint sociobiological metaphors of termite and bee communities are a strange context indeed for first genuine attempts at learning from history. The way the German Physical Society developed and the appearance of the Mainau and Göttingen manifestos against nuclear armament during the 1950s demonstrate that the path actually chosen did in the end lead in the right direction.[346] Although these

---

[344] Ibid. See also Regener's letter to O. Hahn, 18 June 1954 (UAS, SN 16/4) about common responsibility.

[345] "Mitverantwortlichkeit der wissenschaftlich Tätigen"; see Regener [1947]. Compare Regener's draft from 1946, DPGA, no. 40199, esp. Article 3: "The DPG obligates itself and each of its members to support the freedom, truthfulness, and dignity of science and furthermore to stay constantly aware that workers in science have an especially high degree of responsibility for the shape of public life."

[346] The Göttingen manifesto against arming the *Bundeswehr* with nuclear weapons was issued by 18 physicists at Göttingen: see *30 Jahre Göttinger Erklärung. Nachdenken über die Rolle des Wissenschaftlers*

manifestos have frequently been faulted for their apologetic purpose, they did have the positive attribute of making a clean break with the 19th-century apolitical self-image of scientists.[347] Their role models were physicists in the United States, who as builders of the first atomic bomb suddenly became only too well aware of the burden of social responsibility.[348] A letter to James Franck by one of the younger signers of the Göttingen manifesto shows how strong a model Franck's politically responsible actions continued to be after the war. Learning about the plan to drop an atomic bomb in Japan while working for the Manhattan Project in the Chicago Metallurgical Laboratories, James Franck had drawn up a memo-randum suggesting a public demonstration of the bomb in uninhabited territory and urging "efficient international organization for peace" and an "international agreement barring a nuclear armaments race."[349]

In 1957, the experimental nuclear physicist Heinz Maier-Leibnitz (1911–2000) wrote to his role model, Franck:[350]

> Surely you have heard about our step concerning atomic bombs for Ger-many. In doing so I often thought of you. The responsibility we still bear today is, of course, small against what you took upon yourself in 1945. But we wanted at least to make a contribution within our area; maybe developments will prove us right.

The *Physikalische Blätter* also played a part in this development. Werner Heisen-berg's retrospective letter from 1963 to its long-time editor emphasizes its influ-ence:[351]

---

in der Gesellschaft, published by Friedensinitiative Garchinger Naturwissenschaftler, Munich, 1987. The Mainau manifesto, drafted by all 16 participants of the annual convention of Nobel laureates in Lindau in 1955, called for exclusively peaceful use of nuclear energy. It was mainly formulated by Max Born and C. F. von Weizsäcker: See Kant [2002], pp. 21–40, esp. pp. 27–30.

[347] Another example of a constructive route toward acknowledgment of responsibility is the article "Der Eid des Homo Sapiens" in *PB* **2** (1946) issue no. 2, p. 1. Brüche published this article at the suggestion of the American anthropologist Gene Weltfish, placing it at the beginning of the issue.

[348] Besides the so-called Franck Report (see next footnote) and Niels Bohr's memoranda see, e.g., Badash [2003] and Oppenheimer [1948] col. 3: "Without tarnishing the wisdom and far-sightedness of our leading statesmen during the war, physicists could not ward off a feeling of direct personal responsibility for having inspired, undertaken, and finally succeeded to a considerable degree in inventing atomic weapons."

[349] See Lemmerich [2002] and http://www.dannen.com/decision/franck.html for a transcription of the original 16-page document from the US National Archives in Washington, D.C.

[350] H. Maier-Leibnitz to J. Franck, 21 Dec. 1957, Franck papers, RLUC, box 5, folder 3. Cf. von Weizsäcker's correspondence with Eugene Rabinowitch in 1957, among his papers, RLUC, box 9, folder 8.

[351] W. Heisenberg to E. Brüche, 31 Oct. 1963, BLM, folder no. 2.

You also write that I was negatively disposed toward your periodical and I really would like to correct this opinion a little. During the war I considered the periodical important and useful because it could convey to people who had no sense for science and technology at all a certain understanding for this area of life. After the war I hoped that the political influence of the groups who knew what science is would augment enough to make the introductory task along the lines of your periodical unnecessary. Perhaps I was mistaken on this point, however [...]. On the whole, I find that our younger colleagues make too little effort to inform the public about the substance and importance of their research, which in a democracy is a precondition, of course, for attracting public support for science. So if your periodical assists in this regard, I can only welcome it.

Responsibility discourse appeared after 1945 in declarations of intent, prefaces, and, as we have just seen, even in the by-laws of institutions and organizations like the German Physical Society. The special responsibility of a scientist is still a live issue today. The necessity "of substituting the former scholar type, entirely wrapped up in his special area of expertise, with a more generally communicative educator of the young and of raising awareness of his own responsibility with respect to cultural issues" is nevertheless a pertinent insight.[352] Directly after 1945 there was a similarly strong impetus in the East and West of divided Germany to organize events of general educative value. At Göttingen, for instance, Carl Friedrich von Weizsäcker's lectures on "The history of Nature" became extremely well attended. In the introduction to the printed version, he explained why he had chosen such a sweeping topic:[353]

> Lectures with very general content for students from all the Faculties find much success nowadays. Evidently, there is a demand for them. Where does this demand come from?

---

[352]Hahn [1949] p. 4. He quotes there from article 2 of the new by-laws of the regional Physical Society of Württemberg-Baden (cf. the quote on p. 140 above). The editor of *GUZ* **4** (1949) no. 18, p. 9 also attributes to Otto Hahn "a true political sense of responsibility" and accepts the interpretation of science as a "factor of political power." He regarded the founding of the German Research Council (DFR) as an "important sign of an active sense of responsibility" (p. 4). Other examples of such responsibility discourse include the article on "Archimedes and his circles," *PB* **2** (1946), issue no. 4, pp. 73–74 and von Laue [1949a] p. 12, last column.

[353]Quoted from the version published two years later: C. F. von Weizsäcker [1948] p. 5 (its 4th edition appeared as early as 1949). Cf. Hentschel & Rammer [2001] p. 196. Cathryn Carson's contribution in Hoffmann (ed.) [2003] pp. 73–85 provides a sociocultural analysis of postwar educational literature on physics as a consumer product, and Beyler [1996] documents the widespread complaint about over-specialization.

The danger lurking in the specialization of the sciences is being felt more and more acutely. The boundaries erected between the subjects have become oppressive. Science specialties are incapable of giving us a worldview that could offer us any firm support in our bewildering lives. That is why synthesis is being sought, the greater picture is wanted.

At the reopening ceremony of the *Albert-Ludwigs-Universität* in Freiburg on 17 September 1945, various speakers emphasized that the university had to be a community of students and teachers, not a hive of pure specialists. Better basic education in history was also a part of this new scheme: "We need good but also generally educated academics, who are knowledgeable about the history of mankind and not gullible to any cheap and banal historical interpretations."[354] Numerous universities in the western occupation zones offered courses within the broad-ranging *Studium generale* program, while East Berlin, for instance, opted for the lecture on "Political and social understanding of the present," obligatory for all students. This promising start in reforming the university curriculum unfortunately lost steam. The early lectures in East Berlin evolved into the later so notorious politically spiked "Basic studies in the social sciences" and in the West they deteriorated into a superficial and self-serving reenactment of a long bygone *universitas*.[355]

---

[354]Rector Sigurd Janssen's speech at the inaugural celebration, quote from Fassnacht [2000] p. 62; see also pp. 192ff. Max von Laue offered the first lectures on the history of physics at the University of Göttingen. Others later emulated him in his field; Willy Hartner did likewise in Frankfurt.

[355]Rammer [2004] discusses the curricular reform in the direction of *Studium generale* at Göttingen. Laitko's contribution in vom Bruch & Kaderas (eds.) [2002] pp. 382f. mentions the parallel reforms at the university in East Berlin.

14

WORKAHOLISM: "IF WE WANT TO LIVE, WE MUST REBUILD"

Everyone probably immediately associates the term "postwar Germany" with the reconstruction and the resulting economic boom, the *Wirtschaftswunder*. The rebuilding mentality only gained full force in the Federal Republic of the 1950s and 1960s. Many texts nevertheless traced early on the immense significance of this reconstruction motif, in metaphorical form.

In the very first postwar issue of the *Physikalische Blätter*, its editor and manager already prophesied:[356]

> Active hope is possible, [however, ...] against which it is no small consolation that the intellectual foundations have been liberated again as building ground by the downfall of despotism. Those not sharing this hope will notice with skeptical wonder the energetic interventions even now perceptible at many places, efforts to mend torn threads and lay new foundations.

What had happened could not simply be denied and repressed. Something else had to take its place, a contrasting image to give people new perspectives, hopes, and visions for the future. One could hardly imagine a more suitable theme for this than reconstruction. Without any clearing efforts or rebuilding, continuation of research or teaching was anyway unthinkable. In many places, students with political records were often set the prerequisite of "committing themselves to doing reconstruction work" for one or two semesters without compensation.[357] Elsewhere this even became a general precondition for enrollment. The Mitscherlichs observed a "dogged determination with which the removal of the ruins were immediately begun." They diagnosed its "manic streak" as a defense mechanism against melancholy, grief, and feelings of guilt.[358] The biographer of Hans Kopfermann noted very appropriately a collective tendency to mindlessness (*Besinnungslosigkeit*). In a private letter Kopfermann mentioned how he combatted the feeling of despair with artificial optimism and workaholism:[359]

---

[356]"Zur Einführung," *PB* **2** (1946) p. 2. Heinemann discusses the concept of rebuilding at the KWG/MPG also among the British. This concept among their rivals in the eastern sector is discussed in Vierhaus & Vom Brocke (eds.) [1990] pp. 421ff.

[357]The University of Munich is an example, according to *PB* **2** (1946), issue no. 2, pp. 67f.

[358]Mitscherlich & Mitscherlich [1967] p. 40.

[359]H. Kopfermann to Charlotte Gmelin, quoted according to Schlüpmann (forthcoming) p. 388.

> My wife is almost in despair, even though our chances at the moment,
> physically and mentally, are actually quite good and it takes a lot of optimism
> on my part to pull her out of a total physical and mental exhaustion. I myself
> cannot be put down, of course, and believe strongly in the future. You're
> lucky, if you can do that. It makes you virtually possessed with working.

Even victims of National Socialism were subject to this psychological compulsion
to concentrate on working for the future in order to come to terms with the
current hardship. Hartmut Kallmann, who had been dismissed from his position
as department head at Haber's institute in 1933, received an appointment as
extraordinary professor at the Technical University in Berlin after the war along
with his former position at the Kaiser Wilhelm Institute of Physical Chemistry.
Reporting to his former colleague at the institute about the state of affairs in
mid-1946, Kallmann also touched on the sense of scientific isolation:[360]

> We have now officially applied for a research permit and hope that we will
> receive it soon. You cannot imagine how ravenous we are to be able to be
> scientifically employed again; and I think this is the only thing that is left to
> us in this desolate situation, really working on something useful again and
> I think we can, in fact. I would very much welcome being able to visit you
> soon again in more active relations, because we feel very much abandoned
> in that regard. Those terrible years of horror did not go by me, in particular,
> without leaving a trace, of course, and I know precisely what is the only
> thing of importance, namely, really doing some proper work in peace.

In 1947 the former chairman of the German Physical Society, Carl Ramsauer,
wrote:[361]

> The time of need in National Socialism also had a good side to it for
> the German Physical Society. We collected ourselves and discovered the
> foundation upon which we can and intend to rebuild German physics.

This reconstruction motif reappeared in a different light in an engineer's talk
before the international conference on engineering training in Darmstadt that
same year. The abbreviated transcript, published in the *Physikalische Blätter*, shifts

---

Cf. ibid., p. 390: "Whoever dropped his hands into his lap could not hope to survive, of course.
The widespread need, e.g., re. the food situation, that reached its low point in March 1946, generally
motivated a 'Let's get started!' attitude."

[360] H. Kallmann to M. Polányi, 22 May 1946, Polányi papers, RLUC, box 5, folder 2. This passage
continues with a complaint about the all-too-common condition of self-pity, see here p. 130.

[361] Ramsauer [1947] p. 114. A conversation between the chemist Adolf Butenandt (1903–1995)
and Heisenberg strikes a similar note; see Heisenberg [1969] p. 9, Schüring [2006] pp. 257f.

the perspective beyond national borders. Research is regarded as "the most important champion and advocate of a union of European states."[362]—This was not far from the truth, either, as such international joint efforts as CERN and DESY exemplify in historical retrospect. The champion motif stands out particularly prominently in Heisenberg's first letter to Sommerfeld right after the war. Freshly returned from his internment at Farm Hall, Heisenberg averred that physicists in particular have a leading role to play in the supranational struggle against the "lunacy of nationalism":[363]

> for the time being American policy still is following the line: No science or technology in Germany. Scientists over there have been arguing the opposite thesis: International collaboration among scientists, support of German science as a counterbalance to cultural decomposition and proletarization, active engagement to establish order in Europe. I was very happy to see that this time, physicists internationally have stood the test: no one I met there was suffering from the lunacy of nationalism; they were all just as nice to us as before the war, even our Jewish colleagues; and physicists in America are fighting against the nationalists with extreme stridency. When an American general had the cyclotron of Nishima in Tokyo dynamited, four other nuclear physicists in the USA [...] issued a public statement: the general concerned should be brought before a war tribunal as a war criminal because his action was senseless destruction of cultural property; thus the Americans had fallen down to the same level as the Nazis. This statement also shows, of course, that the physicists did *not* manage to carry their point along normal avenues; but I believe that the best among them certainly do have a chance of prevailing over the nationalists in the long run. But for now the benighted wave that Hitler had set off is still traveling around the whole world.

A nationalistic anti-American undertone still unintentionally shines through this avowal against the "benighted wave," tinctured by Heisenberg's idiosyncratic elitism about the special leadership role of "the best." He continued: "I would think that in the long run such an enormously efficient nation is also capable of producing a political leadership that will treat the problems in Europe properly."[364] Even in the postwar constellation such an appeal to common interests within the

---

[362]See H. Klumb: "Naturwissenschaft, Technik und europäischer Aufbau," *PB* **3** (1947), issue no. 7, pp. 209–211.

[363]W. Heisenberg to A. Sommerfeld, 4 Feb. 1946, Sommerfeld papers, DMM 1977-28/136-43 (original emphasis).

[364]Ibid. How up to date this issue still is today, if the words "Iraq" or the "Middle East" are substituted for "Europe"!

physics profession as a whole was able to unite otherwise divided physicists behind a single cause against other social groups, like politicians or the military:

> physicists made the atomic bomb in America, but because of the great political importance of the matter, the military and politicians threw all their energy into isolating scientists from any political influence very early on. The arguments they used [...] were the standard ones: Scientists are international, in part pro-Russian, therefore politically unreliable. The game of intrigue was surely too much for physicists at first.

The road to successful long-term international research was still long and tortuous. The first priority had to be local efforts. Comments from the period indicate how very close some people were to giving up:[365]

> The difficulties in rebuilding German science are immense. I do believe, though, that it is worthwhile joining in the efforts to overcome them at least somewhat over time. One is occasionally tempted to lose all hope; but I think this resignation would be misguided. We have to get through a minimum and then things will eventually get better again.

This incantation to rebuild on new foundations, letting everything else lie buried in the past and radically resorting to a *tabula rasa*, culminates in the following hymnlike refrain:[366]

> If we want to live, we must rebuild, after every last remnant has been torn down. We must start again from the very beginning. A scientist's research is the first precondition for the success of everything. Holy is his mission, and conscious departure on this noble quest makes him artist and herald of what shall be.

Erich Regener speaks with similar single-minded forcefulness on behalf of science in an appeal for more funding for basic research. His "Memorandum regarding the necessity of promoting the physical sciences" was distributed to all the German Ministries on cultural, economic, financial, and internal affairs and was even handed out to state representatives, members of the upper house of parliament,

---

[365] Otto Hahn to U. Martius, 18 Feb. 1947, AMPG, II. div., rep. 14A, no. 2726.

[366] Dr. Hüttner, Werdau i.Sa.: "Das wissenschaftliche Buch," *PB* 4 (1948) p. 267: "Wenn wir leben wollen, müssen wir aufbauen, nachdem alles restlos niedergerissen worden ist. Wir müssen von vorn, ganz von vorn beginnen. Die Arbeit des Naturwissenschaftlers ist die erste Voraussetzung alles Gelingens. Heilig ist seine Aufgabe, und das Bewußtsein der Größe dieser Aufgabe lassen ihn zum Künstler und Künder des Kommenden werden." This plea for generally accessible and popularly comprehensible texts quotes Schiller as its crowning conclusion.

labor union leaders, and political officials. It had received the wholehearted sanction of the Physical Society of Württemberg and Baden,[367] arguing that:

> today's physics will be tomorrow's technology, so it should actually be self-evident that even in hard times promotion of the physical sciences be regarded as a matter of great priority and that, above all, no cuts be made in existing grants. Instead, at all university institutes, both after the end of the war and especially after the currency reform, we see budget cuts, staff dismissals, incidentally vacated chairs not being reoccupied, and much more.

Since, he continued, nowadays any politician must take into account that "the knowledge reaped from physics would be applied to technical and medical problems, and would touch all areas of human activity to a degree hitherto unknown," and since "enormous sums are being expended on physical research" abroad, it would not suffice to restore pecuniary support to "prewar conditions." Funding would have to be "augmented sharply" beyond that, in accord with the rapid development of physics, "if analysis is not to be left behind in a primitive state."[368] The stridency of this appeal, insinuating the new general importance physics played, was effective. Support by policy-makers turned nuclear and particle physics into booming fields of research during the postwar period.

Social psychologists have analyzed this connection between a rebuilding fixation and forgetting. It is barely veiled in passages like Otto Hahn's statement in his newspaper article from 1949:[369] "the western zones are overcrowded with refugees, the worry about building an existence in human dignity overshadows all former prejudices." The verb chosen is "overshadows," not "dissolves," for instance, or "eliminates." Former prejudices (whatever they may have been against) persisted. They were simply covered over. What lay underneath was buried in the obscurity of the forgotten.

---

[367] See Regener [1949]. The typed carbon copy of the original memorandum is available among the Sascha Mogun papers, UAS, SN 26. It was unanimously passed during the administrative meeting of the Physikalische Gesellschaft in Württemberg-Baden in Heidelberg on 29 January 1949.

[368] Ibid. Regener use the same arguments at the local level, in his letter dated 21 Sep. 1951 to the rector of the Stuttgart Polytechnic about a budget hike for 1952 (UAS, SN 26/36) as well as in a memorandum on the reorganization of physics courses at the polytechnic (SN 16/35). The rector, Erwin Fues, employed similarly bold rhetoric in his address on the occasion of Regener's 70th birthday on 15 Dec. 1951 (UAS, SN 16/24, p. 2).

[369] Hahn [1949] p. 3.

15

SIDE-LINING OF EMIGRÉS AND CRITICS

The Zurich historian and philosopher Hermann Lübbe advanced the thesis in 1983 that after 1945 an "asymmetrical discretion" caused victims of the Nazi regime to forgive perpetrators (e.g., the persons who had dismissed them in 1933). According to Lübbe, they kept quiet about these commissions and omissions in order to regain access to the world of German scholarship.[370] This pattern did *not* match detailed analyses of specific examples at the University of Göttingen by Anikó Szabó and Gerhard Rammer, however. The case of the applied physicist Kurt Hohenemser, whose reincorporation among the university staff after 1945 was actively prevented, rather indicate the opposite.[371] This has also been confirmed in a case from the University of Tübingen. Walter Jens described it as "Jungle combat, full of twilight stratagems, defined by arguments that in view of the occupying power could not be put openly on the table [...:] Obstructionism [...] delaying, [...] absurd counter-proposals."[372] With only very few exceptions, scientists dismissed and driven away in 1933 because of Jewish descent or for political reasons remained unwelcome even *after* 1945. Their former colleagues at the scientific institutions did not look forward to their returning. Non-Jewish emigrés who had left Germany for purely political reasons were also regarded with suspicion. The mathematician and later historian of science Otto Neugebauer (1899–1990), for instance, was even rebuffed as a deserter. The dean of the Faculty of the Mathematical and Natural Sciences at Göttingen, the aerodynamicist Walter Tollmien (1900–1968), argued, for instance, that such cases were not entitled to reparations (*Wiedergutmachung*, another postwar euphemism) because they had not been under threat during the Nazi period but had fled into the

---

[370]See Lübbe [1983]. Cf. Szabó [2000] p. 510, however, for a more likely variant of this thesis. Accordingly this discretion was practiced not between victim and perpetrator but between politically implicated professor and fellow traveler, because both sides knew perfectly well the compromises they had each made.

[371]For details see Hentschel & Rammer [2000] pp. 728–739, [2001] pp. 202–206 and Rammer [2004] chap. 5. Schüring [2006] pp. 203f. presents analogous examples of former KWG directors continuing to maintain a distance from their ousted colleagues even after 1945.

[372]Jens [1977] pp. 344f. Cf. the similar findings by Sylvia Palatschek (in vom Bruch & Kaderas (eds.) [2002] p. 402): "Leaving the past alone and the emigrants abroad was the maxim of university politics during the period of reconstruction."

comfortable security of a foreign country.[373] Richard Courant had to contend with similar reasoning by a Göttingen dean, whose maudlin response to Courant's inquiry in 1950 was that Courant's recently acquired American citizenship posed an unfortunate obstacle to his receiving any payments as emeritus of the university. Courant snapped back crossly: "I am not filing any claims and merely wrote to find out what my former Faculty was planning to do in my case. Does the curator of the University of Göttingen really not know that my German citizenship was swiped by the National Socialist government, and do you really believe that I am willing to give up my American citizenship?"[374]

As Anikó Szabó has shown, these weak excuses grew out of a distorted picture of the situation of emigrants (as a life of leisure in sunny California) combined with the residual effects of Nazi propaganda (that had denounced them as defectors) and a more pressing concern for refugee colleagues closer to home who had lost everything in the bombings.[375] Annette Vogt pointed out that unconsciously adopted phraseology minimizing the dismissals since 1933 as "accepting an appointment" elsewhere or as "going into retirement" also discouraged the exiles from putting much effort into exploring an eventual return to Germany. According to more recent estimates, scarcely more than 10–15 percent of emigré scholars from virtually every discipline returned.[376] And the few persecuted scientists who had stayed in Germany, like Ursula Martius, were likewise ostracized (as "soilers of the nest") for publicly criticizing the situation after the war. She was left no alternative to leaving postwar Germany. Wilhelm Hanle described a particularly dire instance of twofold punishment, after already suffering under the Third Reich:[377]

> I know [...] a colleague who was in jail for 10 years for high treason and has tried in vain to return now to the academic profession. I had taken him in after he had been released from jail at my institute as supernumerary assistant, temporarily, because his field was not ours. I don't dare to advise him to follow a university career now in this overcrowded and difficult

---

[373]The Neugebauer case is discussed in Szabó [2000] pp. 445–454, 494f.

[374]See ibid., pp. 420f. about Courant's argument with the astronomer Paul ten Bruggencate (1901–1961). Similar correspondence by expelled KWG members is quoted by Schüring [2006] pp. 170f.

[375]See Szabó [2000] p. 508. Cf. Mitscherlich & Mitscherlich [1967] pp. 65ff. on "emigration as a fault" as late as the mid-1960s with reference to Willi Brandt's candidacy for parliamentary election in the Federal Republic of Germany.

[376]According to Claus-Dieter Krohn in vom Bruch & Kaderas (eds.) [2002] pp. 443, 445ff. and Palatschek's contribution there, pp. 402f.; furthermore, Schirrmacher [2005a] on the problematic concept of "remigration."

[377]W. Hanle to E. Brüche, 18 March 1947, BLM, box 119, folder 308.

situation after so many years. He has lost 10 years and he cannot retrieve them. I had never seen such a blatant contrast between those politically favoritized and those impeded by the Nazi regime as in this case; but it exists on the small scale in many cases. If someone with so few prospects under the future and current difficult circumstances in Germany uses the chance now to go abroad, one shouldn't prevent him, but rather be glad about it.

This hopeless alienation is detectible in Martius's correspondence with Hahn and von Laue from as early as 1947. She tried in vain to explain it to these two physicists, whom she had always admired for their ethics:[378]

> what I have been trying to express is: the feeling of great—actually hopeless—worry. (I would much prefer to be out of this country and be rid of this whole nightmare. It is not so bad that things are going badly and are imperfect, just that people who want to do it differently are basically not wanted because they are troublesome to have around—both from the German perspective as well as from the extra-German one.)

Even formerly ostracized persons who had received personal invitations to return after the war—despite existing reservations against emigrés in general—immediately sensed the alienation. Lise Meitner explicitly mentioned it as the reason for declining an offer to work in the reopened Max Planck Institute of Chemistry in Mainz.[379]

> I have little fear of the unfavorable living conditions but very considerable reservations about the intellectual mentality. All cases where I might be of a different opinion to my coworkers—outside of physics—would surely be met with the words: She doesn't understand the German situation, of course, because she is an Austrian, or because she is of Jewish origin.

---

[378]U. Martius to M. von Laue, 25 Oct. 1947, DPGA, no. 40048. Cf. Rammer in Hoffmann & Walker (eds.) [2007]. The writer Irmgard Keun's letters to Hermann Kesten after her return from exile express a similar mood: see Engelmann [1987], the chapter on "I feel so alien and lost here."

[379]L. Meitner to Otto Hahn, 6 June 1948, quoted from Krafft [1981] p. 185. Cf. an excerpt from Meitner's letter to Franck's daughter Lisa Lisco from 29 May 1955 in Vogt [2002] pp. 126f. Sime [1996] pp. 356f. quotes Franck's similar refusal to Heidelberg: "misunderstandings would arise from differences in experience alone." See also Lemmerich in Hoffmann (ed.) [2003] pp. 191–196, about the reasons for Meitner's refusal of the offer to direct a new Max Planck Institute in Mainz and the impossibility of true reparations. Carl Neuberg also criticized Germany for its lack of will to make amends ("kein Wiedergutmachungswille"). His letter to Adolf Butenandt from 7 Mar. 1947 is quoted in Schüring [2006] p. 148.

Hans Bethe (1906–2005) chose a quite similar tone in refusing the prestigeous chair for theoretical physics at Munich in 1947, having also been dismissed in 1933. Bethe replied to his teacher Arnold Sommerfeld:[380]

> It is not possible for those of us who have been driven out of our positions in Germany to forget. The students of 1933 did not want to attend our courses on theoretical physics (and it was a large group of students, perhaps even the majority), and even if the students of 1947 think otherwise, I cannot trust them.

Erich Regener set as a condition for his return to the Stuttgart Polytechnic as full professor of physics "that I not have to work together either at the Physics Institute or at the University with men who had been involved in my dismissal at the time," adding the optimistic note: "But I may surely assume that this will result of itself from the current situation." He clearly underestimated the astonishing continuity within the civil service, not just at Stuttgart.[381]

---

[380] H. Bethe to A. Sommerfeld, 20 May 1947, quoted from Rammer [2004] pp. 112f., which also reports on p. 479 about student-organized disruptions of Willy Prager's lectures at Karlsruhe in 1933.

[381] Foregoing quotes from Regener's letter to rector R. Grammel, 12 July 1945, UAS 57/177 (personnel file on Regener). Ousted coworkers at the KWIs expressed similar wishes of "not encountering at the start [...] former National Socialists who had been involved in the political persecutions against me at the time"; see Schüring [2006] p. 242.

16

INSENSITIVITY IN COMMUNICATING WITH EMIGRÉS

Ute Deichmann has already touched on this subject in her analysis of the postwar correspondence between German Jewish emigrés from the fields of chemistry and biochemistry and their colleagues in Germany. Anikó Szabó and Annette Vogt have examined the treatment of exiles from Göttingen and from the Kaiser Wilhelm Society. Thus my account may confine itself to a few select examples from among the correspondence of physicists. A conspicuous characteristic of the first letters after the war by physicists inside Germany is a great lack of compassion for the fate of their outcast colleagues. Contrasted against this is their effusiveness in singing their own sorrows to the exiles, shoving any reflections about the past into the background as mere afterthoughts. Needless to say, many of these reports were motivated by an unspoken hope for a coveted CARE package.[382] For instance, Pohl wrote to Born in Edinburgh as early as August 1945, thus even before normal postal communications were reestablished, thanks to an obliging Allied officer:[383]

> Dear Born, a kind offer by Mr. F. A. Johns gives me the opportunity to write you a few lines from Göttingen. [...] Göttingen has become the place of refuge for all people looking for work and lodgings. You would have a good laugh if you saw what things occupy us Göttingers now. But despite it all: we thank our lucky stars that we got this far, and a positive differential quotient is gradually coming into view; albeit we all did imagine the minimum of the curve would be considerably less flat.

The physicist Rudolf Ladenburg was organizing a systematic distribution of CARE packages to scholars in Germany by arranging sponsorships by emigrés. Ladenburg himself spent more than 1,500 dollars over a period of fifteen months on such packages. The letters by the wives of recipient physicists fill a chapter of their own, reflecting the domestic aspect of life in postwar Germany. Marga Planck's (1882–1949) letter to James Franck of 8 Dec. 1946 is just one example:[384]

---

[382]The charitable organization *Cooperative for American Relief Everywhere* (CARE) was coordinating the shipment of packages of staple goods for the needy abroad.

[383]R. W. Pohl to M. Born, 13 Aug. 1945, SBPK, Born papers, 591 (I thank Gerhard Rammer for pointing out this letter to me).

[384]Franck papers, RLUC, box 6, folder 4.

I feel quite horridly demanding. No, your "Care" package had not been pillaged, everything was in it, and it is altogether wonderful. But it seemed so unreasonable to me that for 44 pounds weight there was a content of c. 27 pounds incl. tins and, besides, we heard that these "Care" packages were so terribly expensive! But then the Americans told us that other packages were even more expensive! and that "Care" was therefore not that disadvantageous at all for the sender. It is a terrible thought that you have spent so much money on us—and then, more wishes are sent afterwards as thanks!

By the end of 1947, however, Ladenburg had to observe a significant drop in the number of willing sponsors, owing to the incongruous reactions they received.[385] The following inquiry by the president of the Göttingen Academy of Sciences at the end of 1946 is symptomatic of this insensitivity. Rudolf Smend (1882–1975), professor of public law, plied for permission to reenroll former *membra non grata* of the Academy of Science's Physical Class in a manner that could only be described as presumptuous:[386]

On 21 August 1945 we sent you and a number of other members of our Academy the following letter: Now that the deplorable circumstances that led to your cancellation of membership no longer exist, we request permission to list you as members again. The majority of these letters have evidently got lost. We thus repeat our request that you please consider yourself as a member in the described sense. To our great gratification Messrs. Misch and Latte have long since rejoined our ranks and are participating here again.

By watering down the injustices of the Nazi era and ignoring any personal involvement, this letter could only incense the addressees. Franck wrote to Born:[387]

---

[385] See Schüring [2006] p. 303, based on a letter by Ladenburg to Meitner dated 17 Dec. 1947.

[386] This letter to Franck, quoted in Szabó [2000] p. 511, is also discussed in its historical context in Lemmerich [2002] p. 158. It was considerably more difficult for German-speaking scholars in the humanities to shun such credentials and pursue new careers abroad. The philosopher Georg Misch (1878–1965) and the classical philologist Kurt Latte (1891–1964) had both been forced into early retirement in 1935. Misch was in exile in Great Britain until 1946, while the younger Latte was able to return to his chair at Göttingen in 1945; cf. Szabó [2000] pp. 341, 108. Other similarly euphemistically worded statements by Smend are cited there as well. E.g., in May 1945 he greeted the return of dismissed professors who had not opted for exile as "a fortunate consequence of the abolition of a part of the current law." By contrast, the letter sent out by the German Physical Society in 1953 to members dismissed between 1933–38 sounds downright compassionate in requesting they consider themselves its members: see Stefan Wolff in Hoffmann & Walker (eds.) [2007].

[387] Letter dated 8 March 1947. SBPK, Born papers; carbon copy in the Franck papers, RLUC, box 1, folder 7. Cf. Meitner to Franck, 21 Feb. 1947 and Lemmerich (ed.) [1998] p. 480 and [2002]

> I found Smend's remark simply outrageous, dismissing what had happened in Germany with the words "deplorable circumstances." I can very well picture a group of humanists sitting there who perhaps even consider it a great sacrifice of theirs to ask people like you and me to consider ourselves as their fellow members and that it ought to be an honor for us. I feel I cannot tolerate such an attitude.

Otto Stern (1888–1969) promptly rejected readmission with reference to the terrible events of the past. But Max Born seems not to have had any qualms about becoming an academy member again. Hermann Weyl (1885–1955) also decided in favor of membership after reading a few issues of the *Göttinger Universitätszeitung*, in particular an article by Rudolf Smend about an ecumenical conference in Cambridge: "thus I can be certain that this is not conceived as a lame 'forgive and forget' and that the unanimous and pure will of the Academy to rebuild a truly free, supranational community of researchers lies behind their invitation." Franck shared this view. He wrote in English to Weyl:

> I am not so much concerned about the way in which I was dropped from membership as in the spirit in which the Akademie [*sic*] will work in the future, and I want to make it clear that if I do join again it will be because I regard it as a duty to help but not as an honor.

Even examples of an apparently successful recovery of relations reveal more than an initial fear of reproach and embarrassing questions. There was a quite astonishing insensitivity toward subtler attitudes. By avoiding any attempt to step into the other party's shoes, German physicists evaded painful blushing in the re-opening of former ties after the war. But it is questionable whether their colleagues in exile really were so nonchalant about how the Germans perceived them. More intimate correspondence between emigrés indicates otherwise. Rudolf Ladenburg's letter to Lise Meitner demonstrates how shocked he was with what he had found:[388]

---

about Meitner's indecision about how to respond to Smend's letter. Stern rejected it outright: see Stern to Franck, 20 March and 31 May 1947, both among the Franck papers, RLUC, box 9, folder 2. The positions taken by Courant, Weyl, and Franck in this academy matter are the subject of Franck's correspondence with Weyl (including carbon copies of Weyl's exchange with Courant), Jan.–March 1947, RLUC, box 10, folder 5 (the following quotes are from Weyl's letter to Smend of 18 Feb. 1947 and from Franck's letter to Weyl from 11 Feb. 1947).

[388] Else and R. Ladenburg to L. Meitner, 15 or 12 Dec. 1947, Meitner Papers, Churchill College, Cambridge, vol. 44–48. Ladenburg's sources included, e.g., Courant and the wife of the mathematician Artin (I am indebted to Jost Lemmerich for this reference).

> Meanwhile we also have heard much, too much from Germany, partly
> from letters, partly from colleagues who had been in G[ermany] shortly
> beforehand, particularly at German universities, and had spoken with many
> closer and more distant acquaintances. [...] These reports agree fully
> with your impressions—there seem to be only very few left with whom
> one can communicate, at least among physicists. [...] Laue is constantly
> being praised as a special exception—that says it all, considering that even
> with *him* you missed a complete meeting of minds [...]. Heisenberg is
> apparently just as nationalistic as he always had been, not to speak of von
> Weizsäcker, Pohl, and others. Mathematicians, or at least some of them,
> seem to be more reasonable. Reports about the revival of anti-Semitism are
> very depressing—Kallmann writes in complete desperation—and the Nazi
> sentiment among the younger generation also. One would want to despair.

Let us leave aside whether these severe judgments about nationalistic mentality
dating from the first few postwar years were warranted or at least partly based
on misunderstanding: The impression made by physicists as a group was certainly not positive (even contrasted against mathematicians). It certainly must
have been strenuous to hide such reservations in face-to-face conversation with
persons directly involved. After accepting Bohr's invitation to Copenhagen, Hans
Kopfermann wrote home to his neighbor in September 1948:[389]

> I arrived here a little anxious because one hears everywhere that the Danes
> are still resentful. I must say that I was most pleasantly disappointed. The
> people are all touching, have astonishingly much sympathy and compassion
> and sit so little on their high horse that one is almost ashamed. It is actually
> as if the 10 years lying in-between were not being counted at all. I meet a
> lot of American colleagues here as well, in part German emigrés, who are
> behaving themselves really positively. Weisskopf said to me: "You're in a
> much worse situation than we, you know. We fell down into the larder."

Others had entirely different experiences, of course, than Kopfermann during
his first post-war encounter with Victor Weisskopf (1908–2002). Brüche anonymously published a letter by a physicist working in the USA in 1948, who had the
following to report:[390]

> The only people who are sometimes pointedly reserved (but not one-
> hundred percent either) are the former German emigrés from the period

---

[389] Hans Kopfermann to Charlotte Gmelin, private archive C. Gmelin, quoted from Schlüpmann
[2002] p. 441. Heisenberg was similarly deaf to reservations even against him personally by colleagues
in Allied countries: see e.g., his letter to Sommerfeld, 5 Feb. 1946, DMM 1977-28/136-43: "they
were all just as nice to us as before the war, even our Jewish colleagues."

[390] *PB* 4 (1948), pp. 268–269. The author may have been W. Finkelnburg.

after 1933, some of whom had suffered severely in the past system and who therefore are not happy to see us here, primarily trying more or less obviously to keep us away from their universities in which they are established.

Simply omitting the dozen gruesome years of the Third Reich from the calculation was clearly wishful thinking on the part of people trying to obliterate a bad conscience. Lise Meitner used even stronger words in her letter to von Laue about the fatal consequences of such insensitivity:[391]

> You presumably know that [Wilhelm] Westphal is in America. He wrote to [James] Franck that he had a very clear conscience and was expecting to obtain a good position, possibly through Franck's contacts. Both statements are mistaken. He owes his clear conscience to the circumstance that he has no conscience at all and has just as few chances for a good position. His academic accomplishments correspond entirely to his personal accomplishments.

Her exchange of letters with Max von Laue up to 1948, which Jost Lemmerich has recently published in full, never reverted to the same warmth of before 1933. Two in June 1948 are typical. They were inspired by the controversy surrounding von Laue's unfortunate article in the *Bulletin of the Atomic Scientists* about Goudsmit's book *Also* (see here p. 121). Meitner wrote in response:[392] "Thank you very much for your letter. It demonstrated to me yet again how difficult, indeed, how impossible it is to evaluate issues by letter that cannot be isolated from attitudes governed by the emotions."

Meitner continued to try to make clear to him why articles like his simply achieved the opposite of what they intended, despite "having been written in most genuine sincerity." Instead of conceding that there had been a lack of moral courage in Germany, von Laue responded with a kind of counter-offensive

---

[391] 31 Jan. 1947, quoted in Lemmerich (ed.) [1998] pp. 479–481, quote on pp. 480f.; cf. p. 25 above.

[392] Meiter to von Laue, 24 June 1948 and von Laue's reply of the 30th, quoted from Lemmerich (ed.) [1998] pp. 517–519. Meitner's correspondence with her close friend the botanist Elisabeth Schiemann suffered similarly. Cf. Sime [1996] p. 352 and Meitner to J. Franck, 22 Feb. 1946 (quoted here on p. 167). When the two long-time friends saw each other again in England in 1947, their two value systems clashed head-on: "Some kind of a wall stood between Elisabeth and me. When I asked her whether it was true that there was so much anti-Semitism in Germany, her answer was: If you ever knew how much anti-Semitism exists in England. And when I responded, that may well be true but that did not prevent that Peierls, for instance, was offered a choice between all the theoretical chairs in England, she said nothing in reply. I find this quite depressing. The attitude of our generation is not very important for Germany's development—but what do the younger ones think?" (Ibid., letter from Stockholm dated 10 July 1947.)

against all foreigners who had maintained diplomatic relations with Germany, for instance, and had come in throngs to attend the Olympic Games in Berlin: "It is undoubtedly correct that sins have been committed by us, but foreign countries had sinned much more because opposition would have been completely safe over there." This volley of reproaches was not conducive to gaining insight into personal complicity, negligible though it was in von Laue's case.

It was a widely held view among German physicists who had loyally stayed behind in the fatherland that only those who had experienced the Nazi system from the inside could evaluate what could and what could not be done in that era. This conviction as good as refused any emigré the right to pass judgment on their actions. At the same time it offered a very convenient explanation for some very severe reproaches by emigrés, which were so hard to square with their own self-justifications. A typical example would be a letter by the nuclear physicist Karl Wirtz (1910–1994) to the Norwegian theoretician Harald Wergeland (1912–1987). Significantly, Wirtz uses the pronoun "we" to describe the collective experience of German physicists with their fellow countrymen abroad:[393]

> For no one, who had not been here, *can* know all the parameters of the problem. Only whoever knows all the parameters can validly judge.
>
> We often noticed that we were not being understood when an attempt was made to express these things. A short while ago Bethe was here. I have the feeling that he understood how things stand here. There were so many other visitors here but he alone seemed to have glimpsed the depths of the problem; only he said a few words of understanding. I did not see Miss Meitner, because I did not go to Göttingen. I gather that this is a very particularly difficult personal problem. I do not know her myself.
>
> One can scarcely resolve all this in writing.

It would seem that such misunderstandings and dissonances in exchanges by letter might easily have been avoided in oral discussion. Wirtz's positive impression of Bethe just as his interpretation of Meitner's harsh criticism merely as a personal problem of her own, would appear to confirm this. But even in the best of situations, reticence and insecurity tipped the balance. Hertha Sponer (1895–1986) reported about a meeting with the astronomers Hans Kienle (1895–1975) and Walter Grotrian together with their wives on the occasion of a visit to Dahlem, Berlin:[394]

---

[393] K. Wirtz to H. Wergeland, 2 Sep. 1948, university library of Trondheim, Wergeland papers, box 1. I thank Dr. Roland Wittje for informing me of this letter.

[394] Handwritten draft of H. Sponer's letter to L. Meitner, Franck papers, RLUC, box 5, folder 5.

[...] a somewhat forced meeting, no one said anything (besides von Laue) about what was bothering him: Among these 4 people, Kienle looked the worst, burnt out, more haggard than before, and somehow wretched. The people are really suffering. If only one could invite them over for a visit sometime so that they could relax a bit again for a short time.

My statement is qualified to "in the best of situations" because such meetings among self-assured temperaments could expose even more embarrassing weaknesses. Meitner's encounter with various physicists in Göttingen in 1948, for instance, turned into "a little mental dance on eggshells":[395]

I felt this particularly with Heisenberg and Weizsäcker. During a lunch at Heisenberg's, that Helferich, Clemens Schaefer, and Bonhoeffer also attended, Heisenberg said, among other things, after Helferich and Schaefer had mentioned what terrible things the Germans had done: Every intellectual revolution unfortunately always brought great atrocities along in its wake. I asked him whether one really could compare the French Revolution with Nazism, which at the beginning at least had started from an intellectual and ethical basis. And whether it wasn't more legitimate to compare Hitlerism with the Napoleonic war of conquest. Heisenberg did not respond directly. With Laue, whom I found to be very nervous, I consciously avoided talking much about politics. But he once said to me very agitatedly, if the military had listened to the intellectuals in 1944, the assassination of Hitler would have been successful. I think that this is a great mistake; one only has to read von Hassell's diary to see how very absolutely passively— not to say more—the whole German bourgeoisie stood as regarded active intervention against Hitler. It was very interesting for me, especially in this respect, when Grüneisen told me that when he tried to mobilize professors to pose resistance to the treatment of their Jewish colleagues in 1933 (he had told me about his intentions at the time), he had received the answer, they certainly would not do so, because "what business was it of theirs?"

The total repression of the Nazi past produced excruciatingly embarrassing performances. When in 1945 Theodore von Kármán questioned his former doctoral advisor about German research in applied aerodynamics, Ludwig Prandtl (1875– 1953) alleged he knew nothing about the National Socialists' crimes and the extensive system of concentration camps. This was even though the V-2 rocket,

---

[395] L. Meitner to J. Franck, 27 June 1948, ibid.; compare Sime [1996] pp. 334, 344f. Solid-state specialist Eduard A. Grüneisen (1877–1949) had been employed at the PTR from 1899 until his appointment as professor of experimental physics at the University of Marburg in 1927. 1932–33 he was guest researcher in Berlin. After the Nazi seizure of power, he retained his professorship but refused to cooperate with the Nazi regime.

for instance, was being assembled not far away in Nordhausen under most inhumane conditions by forced laborers and concentration-camp internees in extensive underground bunker systems. Prandtl rejected any responsibility for this and other crimes. He rather complained about the damage American bombers had done to his home and even seemed seriously to expect that in future the USA would finance his research.[396]

It was rather toward colleagues in countries beyond the phalanx of the victorious four that German physicists felt a closer affinity. An emphasis on shared differences from the Allies was a way to draw Germany out of its psychological isolation. Such, at least, seems to have been Werner Heisenberg's strategy in his letters to the theoretical physicist Harald Wergeland in Norway not long after the end of the war. This former fellow at Heisenberg's Leipzig institute from 1937 to 1939 thus had first-hand insight into the situation in Nazi Germany:[397]

> I think that you can put yourself more easily in our place than our American friends, perhaps. The problems are basically, in some ways, simpler for us than over there in the West. For a number of years, now, it has simply been a matter of survival for us here, i.e., trying with a small group of closest friends to steer the little boat of life through a raging hurricane and lend a hand as steadily as possible without this little boat overturning in the process. This sounds very egoistical; [I am] a drifter on the water as long as I do not see how to [be] able to master the storm itself. Yet I still do not believe that this storm is the fault of a few people, rather that a number of uninhibited or weak people [made a series of mistakes ....] The storm, however, is to us a mental derangement of the masses that is coursing around the whole world. At the rate that the masses are being alienated from Christianity, the world seems senseless to them and in the confusion they all do the most atrocious things. On the whole, the natural instinct for right and wrong is certainly more widespread in the West than in the East or over here, but as regards here, we could perhaps add "still." There is no

---

[396] On the foregoing, see von Kármán [1968] pp. 335f. Alessandra Hool kindly pointed out this passage to me: see also Hool et al. [2003] p. 44 and ibid., pp. 47f. for an excerpt from Max Born's report of Kármán's complaint about Germans during a meeting in Moscow on 15 June 1945: "They do not understand the situation and their own responsibility and asked him whether Carnegie or Rockefeller Foundation could not finance their institutes! They think they lost the war because the scientists were not properly made use of!" A tightly spaced typed transcript of this report in English: "Journey to Russia, 13 Jun.–1 Jul. 1945," that was evidently sent to various emigrés, is located among the Franck papers, RLUC, box 1, folder 7.

[397] W. Heisenberg to H. Wergeland, 29 Sep. 1946, university library of Trondheim, Wergeland papers, box 7. Dr. Roland Wittje kindly pointed out this handwritten letter to me. My readings of illegible words appear in square brackets.

helping it, humble folk then have to try covertly, as yesterday the monks, to pass the true values on to later generations.

As if this pseudo-existentialist effusion had not sufficed, Heisenberg attempted in his next letter a deeper contextualization of the actions of his fellow countrymen in their lonely "battle for survival."[398]

> But I do have the basic feeling that our times are not worse than ten or a hundred or a thousand years ago. One also has to realize that what initially appear to be crimes and direct contagion of the depravity of individual persons is often simply the material form in which that great process of life called natural selection unfolds. Only thus is it comprehensible that even the most idealistic and honest of schemes for improving our world—I am thinking of Christianity, later the Reformation, most recently the attempt to eliminate war—almost always lead to the most horrendous atrocities, against which the targeted evil looks ridiculously small in retrospect. That is why moderation is called for here; and one should be very careful with all plans to make the world better. The only thing that remains at all times is reasonable and decent actions on the small scale; and I sense it as an important sign for the better that now good actions are permitted and acknowledged in the West also in politics.

> At the individual level, the destiny and moral attitude of a nation is naturally very strongly dependent on geography, external living conditions, etc. In present-day Germany the battle for survival is being bitterly waged hand-to-hand. Of course, the struggle for survival is hard in a country like Norway as well: but where you live, Nature is the enemy, and people can join forces; and in such low numbers they can extract their food from the soil. Over here the soil is richer but the population density is so absurdly high that the enemy always is one's neighbor. Hopefully the plans for a European economy, which are now being discussed, will also improve this situation one day.

Those dozen years of Nazi rhetoric about a landless *Volk* had left their trace. Heisenberg was evidently expecting salvation from a grand population plan for the whole of Europe. During his visits as ambassador German scientist to the Netherlands, Norway, and Denmark before 1945, he had argued along similar lines. His hosts could only be shocked at such obvious attempts to distract away from the Nazi crimes committed in what were at that time Germany's occupied territories.[399]

---

[398] W. Heisenberg to H. Wergeland, 27 July 1947, ibid.
[399] See Walker [1992] and the sources quoted there.

Such pseudo-historical philosophizing was not unique to Heisenberg. The same lack of sensitivity and thought about the particular gravity, if not singularity, of the crimes committed in the name of Germany, emerges in other contexts. In 1949 Lise Meiter reported about a gathering of German physicists within the framework of conferences in Basel and Como. Walther Bothe, Heisenberg, and Gentner were among the group of younger physicists:[400]

> A very nice Swiss did say to me, who had personally very much supported inviting the Germans: Each individual German is very nice, but many of them together are hard to take. There is something to this, unfortunately. It seems astonishing to me every time how little the Germans, even the best of them, have learned from the events. Bothe, who certainly had never been a Nazi, presented a theory to me in Basel that the Nazis' atrocities stemmed from the Mongol traits that had come from Russia to Germany and was most distinctly visible particularly among Bavarians,— and Hitler did originate from the Bavarian border. Besides that he told me the American camps for "displaced people" were so terrible that German concentration camps were mere child's play in comparison. I just replied: "How terrible" and ended the conversation.

---

[400]L. Meitner to J. Franck, 19 Oct. 1949, Franck papers, RLUC, box 5, folder 5; compare also Arendt [1993] pp. 29ff. about the "inability and aversion to distinguishing at all between fact and opinion," prevalent among Germans.

17

DISTRUST AND OBDURACY AMONG EMIGRÉS

Victor Weisskopf's open-minded and conciliatory attitude toward Kopfermann was rather an exception among emigrés. Only their younger German colleagues, who were more easily deemed politically innocent, could hope for such treatment or the few individuals like Karl-Friedrich Bonhoeffer (1899–1957) who earned such trust from losing family members involved in the resistance movement to Nazi executioners. James Franck wrote him after the war:[401]

> It gives us confidence in the future to know that people like you continue to exist in Germany, who have kept a clear head in a sea of madness and moral degeneracy and guarded their knowledge of what is right and what is wrong. [...] I understand that it must be hard to explain to your children that all that is happening now is the result of a gangster government. This is a matter of fact but that does not mean that we over here do not find much that is happening in Germany now deplorable and disapprove of it. It is only insofar explicable as Hitler's total war also whipped up so many bad instincts that it will take a while for the whole world to be able to learn more healthy ethical compassion again. This does not make it easier to endure when one sees oneself surrounded by hunger, cold, and misery and not be able to protect one's children and relatives. [...] Many even among the best and dearest of people to me in Germany seem to be losing patience and hope. In such misery that is understandable but not right and I am glad that this is not so with you.

Deep-seated distrust was much more common among former colleagues divided by the prewar events.[402] The opposite extreme, but just as atypical, was Albert

---

[401] J. Franck to K. F. Bonhoeffer, 2 March [no year, approx. 1946], AMPG, div. III, rep. 23, no. 20,9.

[402] Paul Ewald, for instance, attributed Hans Bethe's refusal of Sommerfeld's chair to this incompletely understood "distrust among emigrés": Ewald to A. Sommerfeld, 1 April 1947: www.lrz-muenchen.de/~Sommerfeld/KurzFass/01129.html See also Bethe's own letter to Sommerfeld, ibid., /03746.html Cf. J. Franck to Otto Hahn, 18 Jan. 1947 (Franck papers, RLUC, box 3, folder 10): "You know, of course, how acutely I feel the unjustified distrust that hampers Otto's research. Believe me, I tried many things, apparently with little success until now. I am also very unhappy that all this trouble affects non-Nazis even worse than the Nazis." Other examples of mutual distrust are mentioned by Schüring [2006] pp. 321ff.

Einstein's repudiative stance. From bitter personal experience with authoritarianism and chauvinism, even among his former colleagues in Germany, he deemed it practically impossible to "turn those fellows into honest democrats." (Einstein was one of the rare exceptions among his fellow Germans who would not be swept up in the initial euphoria about World War I and stolidly advocated pacifism.)[403] Add to the mix his abhorrence and disgust with the crimes that had been committed in the name of Germany and the Germans. Einstein felt psychologically compelled to sever all ties with Germany forever: "After the Germans slaughtered my Jewish brothers in Europe, I want to have nothing to do with Germans ever again."[404]

At the end of 1948 Einstein received a letter from Otto Hahn in his capacity as president of the Max Planck Society, founded in the British and American zones in February 1948. According to Hahn, the intention was "to carry on in the tradition of the Kaiser Wilhelm Society before 1933; the society's statutes also have been drawn up with the approval of the American and British military governments like the statutes of the Kaiser Wilhelm Society before the Nazi period." Many former members of the society who had emigrated had already joined as "foreign scientific members" of the renamed Max Planck Society: for instance, James Franck, Otto Meyerhof (1884–1951), Rudolf Ladenburg, and Richard Goldschmidt (1878–1958). The society's senate, the highest administrative organ, included persons of integrity who had been forced to resign their offices in 1933. The former Prussian minister of culture Adolf Grimme was one as well as the former representative of the Center party, Prelate Georg Schreiber (1882–1963), who had been an important science and culture policy-maker during the Weimar Republic. "From these names you can see that any sort of revival of National Socialist tendencies in our new society is excluded."[405] A typical attribute of this letter is the strained attempt to cling onto pre-1933 traditions, completely ignoring what had happened since. Einstein's answer was harsh:

[403] Einstein's complicated political stance is discussed in Nathan & Norden (eds.) [1960], Goenner & Castagnetti [1996]; see also the subsection on pacifism in Grundmann [2005] pp. 254ff.

[404] Einstein, quoted in Hermann [1979] chap. XIV: Einstein und die Deutschen, p. 117. Hermann Broch strikes a similar note in his letter to Volkmar von Zühlsdorff dated 9 Aug. 1945: "The worst thing I went through in Hitler's Germany was a repugnance for people and their stubbornness. [...] I do not want to relive the baseness I personally experienced (at university, etc.) again; it may sound like whining but I don't want to have to see those people again." Broch [1986] p. 26.

[405] Above quotes from O. Hahn to A. Einstein, 18 Dec. 1948, cited in Hermann [1979] p. 117. On the following see a facsimile of Einstein's response dated 28 Jan. 1949, ibid., p. 119, from MPGA, Abt. III, Rep. 14A, No. 814, Bl. 5. The refounding of the KWG/MPG was legally and politically complicated; see, e.g., Manfred Heinemann in: Vierhaus & Vom Brocke (eds.) [1990], Oexle [1994], and the primary sources and further references cited there.

Dear Mr. Hahn, it is painful for me to have to send you, in particular, i.e., one of the few who maintained their rectitude and did their best during those evil years, my refusal. But there is no other way. The Germans' crimes are really the most abhorrent that the history of so-called civilized nations can put forward. The attitude of German intellectuals—regarded as a class—was no better than the mean public. There isn't even any sign of remorse or an honest will to recompense what little there is left after the gigantic killings. Under these conditions I feel an irresistible aversion to participating in anything that embodies a part of German public life, simply out of a need to stay clean.

Einstein even rejected an invitation to attend an event celebrating Max von Laue, "despite my truly deep sympathy and admiration for my old friend." His reason was "after all that has happened in Germany, I have drawn a line that I will never cross for the rest of my life."[406] His curt refusal to support a humanitarian appeal for the German population, who were suffering dire shortages of food, housing, and heating fuel, that James Franck had drafted in December 1945 is consistent with this inflexibility:[407]

I still remember the Germans' "campaign of tears" after the last war far too well to fall for that again. The Germans slaughtered millions of civilians according to a carefully conceived plan in order to steal their places.*[408] They would do it again if they could. The few white ravens among them changes absolutely nothing. From the few letters I have received from there and reports by a few reliable persons who had recently been sent there, I see that among the Germans there isn't a trace of remorse. I also see very clearly that the catering to the Germans has started all over again at the "United Nations"; these trends, the motivation behind the nursing of Germany back to strength after 1918, are most vibrant among the English; for concern about one's precious purse is stronger there, too, than any worries about one's dear fatherland.

[406] Einstein to Brüche, 20 Aug. 1949, BLM, box 1, folder 12.

[407] Einstein to J. Franck, 30 Dec. 1945, Franck papers, RLUC, box 2, folder 7 (published in Lemmerich [1982] p. 142, also excerpted in Szabó [2000] pp. 427f.). See also Broch [1986] pp. 52ff., Lemmerich [2002], pp. 150ff. The impression an Allied Control officer gained about the food situation was: "A Russian officer, familiar with both countries, told a British correspondent during the Potsdam Conference that he thought the Germans looked fatter, less jaded, and better clothed than the British; four months later, after touring Germany, the correspondent was disposed to agree." Balfour & Mair [1956] p. 12. The nutritional situation worsened further, however, in 1946/47; ibid., p. 152: for instance, in May 1947 the number of calories per day per person in Wuppertal was just 650, compared to an average of 1,040 calories consumed in the American zone.

[408]* Einstein's original footnote: "If they had slaughtered you, too, it would certainly not have happened without the shedding of a few proper crocodile tears."

Dear Franck! Keep your hands off this foul affair! After abusing your kind-heartedness, they will make fun of your gullibility.

James Franck's response to Einstein is quite a contrast:[409]

I am of the opinion that an appeal like the one planned will help if it originates from people who could not possibly have ascribed to Nazism. It will do more harm than good, however, if it leads to an open split within the community of immigrants. It might interest you to know that I personally am not the type of person who is willing to forgive all the sins and crimes. I would have wished that the Nazi mob had been cleared out entirely differently than has been done or as is being planned. On the other hand, I have friends in Germany to whom I am attached, not Nazis but men like Laue, Hahn, and Hertz. I do not intend ever to set foot in Germany again, because I do not want to come into contact with people who have said yes to Nazism, but I will have no part in the punishment and gradual elimination of the innocent.

Because Einstein refused to budge from his position and even threatened, if necessary, to speak out against Franck's appeal at some suitable occasion, Franck was deterred from his plan and limited his efforts to sending out CARE packages. It is a known fact that Einstein never again set foot in the "land of mass-murderers."[410] Franck, on the other hand, did ultimately manage to overcome his aversion and return to Germany a number of times. Nevertheless he could not commit himself to returning to Germany permanently when a chair in experimental physics at Heidelberg was offered to him in late summer of 1947. His American citizenship and acclimatization to Chicago were not the only reasons for this decision, as Franck wrote to the ministerial official Thoma:[411]

---

[409] 11 Dec. 1945, Franck papers, RLUC, box 2, folder 7. See also Lemmerich (ed.) [1998] for an excerpt from Lise Meitner's letter to Elisabeth Schiemann from 22 May 1946 about the conflict between a sense of duty to help and the memory of the millions of people who had died in the gas chambers. Broch [1986] p. 53 also refers to a "campaign of tears" in his letter to von Zühlsdorff dated 14 Feb. 1946. He considered it "betrayal to human justice" to join it: "Germans are still better dressed and fed today than [people] in the other countries."

[410] Einstein's opposition to "particularistic nationalism" in Germany is mentioned, for instance, in the Einstein–Born correspondence, as well as in Ladenburg's letter to von Laue, 20 Nov. 1949, M. von Laue papers, AMPG II, div., rep. 50, no. 1158 (cited in Schlüpmann [2002] p. 434). A similar profound hatred of Germans and their culture is expressed in Samuel Goudsmit's letter to Paul Ehrenfest from 1 May 1933, cited in Meyenn (ed.) [1993] p. 291.

[411] Franck papers, RLUC, box 3, folder 11. Quoted in full in Szabó [2000] pp. 428f.; cf. Lemmerich [2002], p. 161.

A second, more personal reason has to be mentioned that makes impossible my acceptance of a professorship in Germany. The majority of Germans, as I believe, repudiated the murder of Jews and other races that the Nazis characterized as inferior, and I do not accuse them for not throwing themselves down the Moloch's throat when they regarded it as useless. Another considerable percentage of the population, however, were indifferent about these crimes. I do not want to come anywhere near this latter segment. I cannot imagine a profitable career as a teacher in which I had to ask myself whether this or that person with whom I had official or personal business had not belonged among this segment of the population and perhaps even still belonged among them at heart. The good Germans will have to deal with such elements by themselves: as a man unwilling to forget his Jewishness, I cannot lend a helping hand.

Moreover, I am convinced that it will be virtually impossible for people who did not live in Germany during the Hitler period and the war to exert a direct influence on education in Germany. Misunderstandings are foreseeable by virtue of the different experiences alone; and as people are only too inclined to count only their own sufferings, they will constantly distrust any returned emigrants, because they did not share the special sufferings of the German population.

This is perhaps the clearest analysis of the underlying reason why communications between emigrés and their fellow physicists inside Germany failed so frequently. It may well have been inevitable. Any letter from former friends in Germany could only bring nasty surprises and personal disappointments. Some retreated from it by cutting all ties with Germany. The exchange of thoughts amongst emigrés intensified correspondingly. With great relief Lise Meitner wrote in her first postwar letter to James Franck upon her return from the United States about being "able to talk with a person for once, whose judgment, sense of justice, and humanity I have unshakable trust in and, based on a common past, speaks the same language as I." Conversely, Meitner was no longer able to share this trust with her German penpals, like her close friend Elisabeth Schiemann. "Shortly before I left Europe I received a long letter from Elisabeth full of lamentations about the destruction in Germany and not a single word about what Germany had done in the subjugated countries, not to speak of what had happened to Germans inside concentration camps. I am very unhappy about these things. I love Germany but I feel like a mother whose favorite child has become ill-bred. The fact that here [in the USA], in my view, most things are going wrong too does not improve matters."[412] This source reflects like no other the profound

---

[412]L. Meitner to J. Franck, Washington, 22 Feb. 1946, RLUC, box 5, folder 5.

conflict between a heartfelt admiration for German culture that emigré Germans had grown up with and an abhorrence for the perversion of the Nazi period. The fact that so few Germans called this cultural barbarity by name even after the war only augmented the clash of values.

Fig. 18: Otto Hahn in deep conversation with Lise Meitner in 1959. Source: Lemmerich [2003] p. 129.

18

THE MENTAL AFTERMATH

Having come to the end of this study, let me step back a bit to gain some perspective and allow the horizon to come into view for comparison against other reference groups—technicians, chemists, mathematicians, scientists, or even the Germans as a whole from that period, perhaps even against entirely different groups from entirely different periods. What was the mental climate of the years from 1945 to 1949? A collapse after a mere dozen years of an empire that had ambitiously forecast a millenium would most immediately be described as a time of crisis. We occasionally encounter this term among our sources (for instance the quote by Max von Laue, on p. 77). From the point of view of the history of mentality, it would consequently be possible to compare this period against other crises, such as in late 19th- and 20th-century physics or the much debated crisis toward the end of the Middle Ages.[413] Closer analysis yields little agreement, however. German postwar society underwent powerful upheavals, but there was neither a complete dissolution of society nor massive displacements in the social roles of leading members of the groups. The initial wave of dismissals was particularly high in the American and Soviet zones. Depending on the degree of politicization and NSDAP membership, a university's staff could be diminished by over a half. But such proportions were soon scaled down again through a variety of appeal procedures. Another strategy frequently used was simply to move into a different zone. According to newer estimates, maximally one quarter of all politically incriminated academics had to seek new occupations. The remainder found their way back into the university system by 1950.[414]

The ideologues in physics had already been pushed aside in the final years of the Nazi regime during the reorganization of physics. The physicists exercizing key functions at the close of the war were thus able to maintain and even build on

[413] On the former see, e.g., Hentschel in A. Nordmann et al. (eds.) *Heinrich Hertz: Classical Physicist, Modern Philosopher,* Dordrecht: Reidel, 1998, pp. 183–224, or Alexei Kozhevnikov's talk about the crisis of quantum theory 1922/24 at the conference on *100 Jahre Quantentheorie in Berlin* held in December 2000. On the latter see, e.g., Klaar [1994] and further references there.

[414] According to Ash in Judt & Ciesla (eds.) [1996] pp. 64–67, a total of 4,298 academics were dismissed after 1945. By 1950, 3,479 of these were still living in Germany and 1,301 (or 37.4%) had already been rehabilitated.

their influence after 1945 (with only a few exceptions like Esau and Schumann). Heisenberg, Gerlach, and Hahn were among this influential elite, seconded by Becker, Brüche, Finkelnburg, Kopfermann, Heckmann, and others.

The loss of norms and values, a typical characteristic of crises, also remained within psychological bounds. Elements of the Nazi ideology could easily be brushed off as unwelcome atraditional impositions. Other components of Nazi mentality that continued to persist went unnoticed by German physicists in post-war Germany. When very young, ostracized, or exiled members of the physics community (e.g., Martius, Hohenemser, or Meitner and Franck) publicly faulted a general lack of coming to terms with past events and demanded more severe consequences for politically implicated leaders, they were shunned as individual "soilers of the nest" and pushed aside with as much vehemence as the scapegoats of Aryan physics had been thrashed. The postwar mental landscape I refer to at the beginning of this study was thus surprisingly uniform, excepting only the three categories just mentioned. This is astonishing when you consider the fractious rivalries among physicists during the Weimar Republic pitting theoreticians against experimental physicists, Berliners against non-Berliners, Prussians against south-erners. The professional homogeneity after 1945 is a striking contrast. The potent unifier within the postwar physics community in Germany was a common sense of injustice. Unequal governance among the Allied occupying forces, debilitating restrictions on research, and bewildering denazification policies fed indignant re-jection of all forms of foreign interference. An effect of "eye-winking solidarity" with politically implicated members was what resulted from the mentality of this union of necessity ("Mentalität der Notgemeinschaft").[415] Many other compo-nents at a lower level of "the German mentality" were also common currency. They include conceptions of the role of women or the laying of nationalistic claim to major historical figures, converting Copernicus, for instance, into a German.[416]

The collective identity of physicists right after 1945 constituted a much stronger national identity than during the Weimar Republic or after 1950. Similar to the case of French chemists in the Third Reich (the subject of a recent study by Ulrike Fell), among physicists the process of identification took place "in the

---

[415]See Rusinek [1998] which takes as an example the success of an impostor, who climbed the professional ladder to the very top of German academia and was officially appointed rector.

[416]The difficulties women encountered in following a career in physics appear in the controversy between Carl Ramsauer and his student Dorothee Lilienthal in *PB* **3** (1947) p. 256 and **5** (1949), pp. 95–97. On Germanic claims to Copernicus, which were typical of Nazi historiography, see Schimank in *PB* **4** (1948) pp. 382f. and the critique by Volker Remmert: "In the service of the Reich," *Science in Context* **14** (2001) pp. 335–359.

tension between self-perception and alienation, between appropriation and delimitation."[417] Whereas the French localized their 19th-century enemies on the other side of the Rhine, Germans perceived the no less alien Allied occupiers as the enemy inside the fatherland. In both cases the effective cement was the sociopsychological mechanism of the common enemy. A general feeling of insecurity and lack of direction are hallmarks of a crisis. They arose less from self-doubt than from a constant uncertainty about what the Allies might or might not sanction, punish, or promote. On one hand, Allied propaganda was proclaiming (e.g., in the form of flyers) since the fall of 1943 "No revenge against the mass of the German people." And somewhat less positively since the summer of 1944 "We have no interest in annihilating the German people" and the frequently broadcast BBC slogan "Better an end to terror than terror without end."[418] On the other hand, the Nazi demagogues had purposefully spread fear of Allied vengeance in their effort to mobilize the last reserves for total war. Well-intentioned attempts by the Allies to rattle the conscience of the German public during the first days of occupation were a clumsy flop. Poster photos of gruesome scenes from the concentration camp in Buchenwald and a superimposed accusing finger pointed directly at the viewer with the legend "You are guilty!" rather achieved the opposite (see also fig. 17).[419] The change of personal identity that roughly 60,000 people opted for in West Germany alone was also driven by a fear of punishment—I leave open whether or not legitimately.[420] Similar to Georges Lefebvre's impressively described *Grande Peur* of 1789, a collective state of fear that spread very rapidly to wide segments of the French population, the fear among Germans in 1945 caused panicky reactions to Allied initiatives and a powerful wave of solidarity among people who would hitherto never have even remotely considered supporting each other.[421]

---

[417] See Fell in Jessen & Vogel (eds.) [2002] pp. 115–142, quote on p. 117.

[418] See, e.g., Balfour & Mair [1956] pp. 19f., 34f., 48ff. and 170ff. for various conflicting statements issued by Morgenthau and Churchill, as well as Allied flyer and radio propaganda for Germany.

[419] Arendt [1993] p. 48 describes this poster and explains why it had this undesirable effect with the argument: "How could they feel guilty if they hadn't even known about it? All they saw was the accusing finger clearly pointing at the wrong person. From this error they concluded that the whole poster was a propaganda lie."

[420] See Rusinek [1998] pp. 19f. An omnipresent, collective state of fear (*la Grande Peur*) is a mentality-molding emotion and one of the important elements of revolutionary sensibility: cf. Lefebvre [1932b] pp. 88ff., Vovelle [1985] pp. 82f., 88ff.

[421] Georges Lefebvre and historians of the *Annales* school identified this as a reason for the birth of bourgeois society in revolutionary France: see Lefebvre [1932b] p. 110 about the new "sense of solidarity" produced by the "great terror." It provides in our case an explanation for the peculiarly hermetic solidarity among erstwhile enemy camps within the German physics

The *Wendezeit* of 1989/90 ending in the unification of divided postwar Germany provides similar parallels: a ubiquity of notorious turncoats and even many an unnatural alliance among the *Ossi* East Germans against the invading *Wessis*, whose arrogance and capitalist superiority, bordering on paternalism, fills the role once played by the Allies in 1945. Accusations of a new form of Wessi "colonialism" in East Germany were countered by complaints about "perpetual whimpering and self-contented offendedness" by the other side.[422] At both epochal turning points, a similar polarization occurred. The opposing camps blocked any attempts at fruitful collaboration. To a considerable extent, the failure of denazification after 1945 was the result of such a counterproductive social polarization, originating from poor decisions taken during the years of transition to a reunified Federal Republic of Germany. Ultimately the chance for a profound reappraisal and far-reaching reforms slipped by, yet again (such as writing a new federal constitution or restructuring the some dozen federal *Länder*). Both sides believed in the political promise of an economic revival, of flowering gardens of plenty; all were blinded by the silvery glint of the western *Deutsche Mark*. The American greenback had sponsored a similar purging of thoughts and conscience in the feverish reconstruction from 1949 on.[423]

To recapitulate, the most prominent features of the mental profile of physicists and other scientists, technicians, and their contemporaries between 1945 and 1949 are: Insecurity and fear (of the Allies) and bitterness, not just in the face of extremely harsh living conditions right after 1945 but often also from a sense of being victimized or punished for crimes others had committed. "Hunger, nervosity, and insecurity" is Brüche's succinct synopsis in his diary entry from October 1945.[424] Added to that is an insensitivity toward the sufferings of others, self-pity, and sentimentality, coupled with cramping narrow-mindedness. It explains the difficulties Germans encountered in communicating with emigrés. It also explains the widespread feeling inside Germany of living in a topsy-turvy world in which nothing was permanent or reliable anymore and in which they were

---

community. The attendant circumstances are also analogous: political upheaval with an unclear future, high unemployment, partial changes in the leading elite, social uprooting, and extreme food shortages and destitution.

[422]Both quotes from Benz [1996] p. 83, who also drew this parallel about "cultural forms of defeat" after 1945.

[423]Not all *Ossis* (are willing to) recognize such parallels. The East German historian of science Dieter Hoffmann does agree with me, however, that such parallels are legimately drawn. Other structural parallels and differences among the turning points in Germany 1933, 1945, and 1989/90 are discussed in Ash [1995a & b], [1999], [2001].

[424]E. Brüche, diary no. IV (BLM, box 7), 10 Oct. 1945.

being judged and sentenced according to incomprehensible norms. An official of the Allied occupying forces wrote in his diary:[425]

> One had the painful realization that even well-bred and supposedly well-educated Germans regarded themselves as essentially right and misunderstood by the world. I found it a fascinating but depressing experience to struggle towards some kind of mutual understanding. If "re-education" is [im]possible even under such favorable circumstances, what can we hope to achieve with the masses?

The sociopsychological antagonism against the Allies also explains the conspicuous and—for Modernity—unique homogeneity of the mentality characterizing these five years.[426] These groupings of physicists did not form from this defining boundary initially but from a common base within the profession, which was still respected above personal advantage. Forming "islands of stability" for the survival of physics beyond the period of National Socialism was the motto by which Max Planck operated in 1933. It was what prompted him to urge his colleagues, like Heisenberg, to stay. The questionable compromises and concessions that this professional policy led to in satisfying the demands of the regime has meanwhile been described in many places.[427] But compared to the Weimar period with its famous trench battles between various physicist camps, these differences disappeared for a short while in the immediate postwar period, as it were as terms of higher order against the backdrop of greater basic conflicts with the "occupiers." Belligerence against the Allies permitted the bridging of internal differences and strengthened the sense of solidarity among physicists as one very active and influential special interest group in science policy. Sincere and effective "denazification" was condemned to failure against this landscape of escapism and repression, a settling of accounts on the backs of others, and a conscious or unconscious inability to feel remorse.

On the other hand, the physics community's active role in science policy of the postwar period also afforded new opportunities. At Göttingen Otto Hahn,

---

[425] Edward Y. Hartshorne in Tent (ed.) [1998] p. 37 (30 April 1945); see also p. 17 (18 April): "one has the feeling of being able to understand these people although their mentality is as foreign as it could possibly be."

[426] See Ariès [1978] p. 422 on the tendency of a mentality to be pulverized into many micro-mentalities of small and smallest social groups. On conflicts within the physics community during the Weimar Republic, see Forman [1974], Hentschel (ed.) & Hentschel (ed. asst./trans.) [1996] pp. lxx ff.

[427] See, e.g., Mehrtens [1994] on "collaborative relationships" during National Socialism in general, Maier (ed.) [2002] on the engineering sciences, as well as Hentschel (ed.) & Hentschel (ed. asst./trans.) [1996] pp. lxv ff. specifically on physics.

Werner Heisenberg, Carl Friedrich von Weizsäcker, and other central figures approached British officers—in the geographical as well as the professional sense—who shared a similar background in physics. Thus they were able to influence the reconstruction plans not just for physics but for the entire infrastructure of scientific research in the western zones. Erich Regener built equally close ties based on trust with Allied control officers in the French and American zones. They included the head of the Intelligence Service of the US Army, Colonel Brunton, and the head of Section technique, Général de Saint Paul in Lindau by Lake Constance.[428] Robert Havemann, Peter Paul Thiessen, and a few other key figures established similarly good relations with Soviet officers within the eastern sector. The export of scientists and engineers to East and West, criticized from so many quarters, surely contributed to the fading of old images of "the enemy" and created new contacts that were of great use in the increasingly internationalized arena of scientific research. Numerous invitations and hirings of German scientists abroad also had a positive effect on the reformation or restoration of scientific and economic relations.[429] The Allies' leniency toward the army of fellow travelers, "denazified" by a combination of whitewash certificates, forgetting, and repression of the past, offered them a second chance rather than permanent marginalization. It was a chance that most knew how to make the most of. Thus the "reconstruction" of physics after 1945 was more than a simple reinstatement of what had been. It was the building of new, more democratic structures. This was paired with a new consciousness of responsibility for the consequences of one's own actions, a legacy of the atomic bomb. This awareness permeates the famous Mainau and Göttingen declarations against atomic weapons for the army of the Federal Republic of Germany along with many other documents written by physicists in the postwar period.

---

[428]See Erwin Schopper in Becker & Quarthal (eds.) [2004] pp. 302–304.

[429]This point is emphasized by Judt & Ciesla (eds.) [1996] p. xiv: Volker Berghahn, ibid., p. 8, described the FIAT program as a kind of "conveyor-belt for future business connections," and Werner Abelshauser, ibid., pp. 107–118, claims "immaterial reparations" were an integral part of a "reintegration of West Germany into the world market."

ACKNOWLEDGMENTS

The national German research fund, the *Deutsche Forschungsgemeinschaft*, generously granted me a stipend for conference travel and archival research. It afforded me the means to peruse the papers of James Franck, Eugene Rabinowitch, and Michael Polányi at the Regenstein Library of the University of Chicago. I thank the archivists there as well as Herr Ralf Hahn at the Archive of the *Deutsche Physikalische Gesellschaft*, Frau Kazemi at the Archive of the *Kaiser-Wilhelm/Max-Planck-Gesellschaft*, Frau Genrich at the *Mannheim Landesmuseum für Technik und Arbeit*, and Dr. Norbert Becker at the University Archive in Stuttgart for their advice and for granting permission to reprint excerpts from documents among their archival holdings. Archivists and staff members at various other museums and archives have also kindly supported my research by providing me with high-resolution images and other resources needed for this book. I would like to mention in particular Dr. Ernst Böhme of the Göttingen city archives, Mrs. Annette Wolpert and Dr. Susan Becker of the BASF corporate archives in Ludwigshafen, Germany, Manfred Mardinskij from the digital picture archive of the *Haus der Geschichte* in Bonn, Mrs. Dorlis Blume M.A., and Anne Dorte Krause from the *Deutsches Historisches Museum* in Berlin. I am also grateful to Michael Grüttner, my wife Ann M. Hentschel, Dieter Hoffmann, Alessandra Hool, Andreas Kleinert, Jürgen Koch, Jost Lemmerich, and Gerhard Rammer for helpful and critical comments on earlier drafts of this book. Ute Deichmann kindly sent me a copy of a hitherto unpublished letter about H. Kallmann. Roland Wittje pointed me to the Wergeland correspondence at the Trondheim University Library in Norway, and Wolfram Pyta to the methodological parallel with Frank-Michael Kuhlemann's study on religious mentalities. I am grateful to Gerd Graßhoff, Chair for the History and Philosophy of Science at the University of Berne, for kindly granting me access to the institute facilities as a foreign stipendiary. Many thanks also to the senior editor of the physical sciences, Sönke Adlung, and the very efficient production team at Oxford University Press.

LIST OF ABBREVIATIONS

AEG Allgemeine Elektrizitäts-Gesellschaft, electrical company

AMPG Archive of the Kaiser-Wilhelm/Max-Planck-Gesellschaft, Berlin-Dahlem

BIOS British Intelligence Objective Subcommittee

BLM Brüche papers, Landesmuseum für Technik, Mannheim

DFG Deutsche Forschungsgemeinschaft: German Research Association (formerly Notgemeinschaft der Deutschen Wissenschaft)

DFR Deutsche Forschungsrat: German Research Council (formerly RFR)

DHM Deutsches Historisches Museum, Berlin

DMM Manuscripts department and library of the Deutsches Museum, Munich

DPG Deutsche Physikalische Gesellschaft: German Physical Society

DPGA Archive of the DPG, Magnus-Haus, Berlin

FIAT Field Information Agency, Technical (USA)

GDR German Democratic Republic (DDR), former East Germany

GUZ *Göttinger Universitäts-Zeitung*, place of publication: Göttingen

JCS (Allied) Joint Chiefs of Staff

KWG Kaiser-Wilhelm-Gesellschaft: research society, renamed MPG after 1945

KWI Kaiser Wilhelm Institutes: research institutions, renamed MPI after 1945

MPG Max-Planck-Gesellschaft: Max Planck Society (formerly KWG)

NSDAP Nationalsozialistische Deutsche Arbeiter-Partei: National Socialist German Workers Party (Nazi party)

NYT *The New York Times*

PB *Physikalische Blätter*, place of publication: Weinheim

PTR Physikalisch-Technische Reichsanstalt: German bureau of standards, Berlin

RFR Reichsforschungsrat: National Socialist Reich Research Council in Berlin

RLUC Manuscript department of the Regenstein Library, University of Chicago

SA Sturmabteilung: Nazi Storm Detachment

SBPK Staatsbibliothek zu Berlin Preußischer Kulturbesitz, Berlin

SD Sicherheitsdienst: Security service during the Nazi period

SMAD Soviet Military Administration of Germany

SS Schutzstaffel: Nazi elite guard

UAG University Archive Göttingen

UAS University Archive Stuttgart

ARCHIVAL COLLECTIONS

Karl-Friedrich **Bonhoeffer** papers, AMPG, III, rep. 23

Ernst **Brüche** papers, BLM, correspondence box nos. 1–9, 90, 104, 119, 123, and the diaries in box 7

Papers of **Bulletin of Atomic Scientists**, RLUC, box 32 with memoranda by Heisenberg and von Laue after 1945

Correspondence about regional associations (*Gauverbände*) matters of the **DPG**'s executive, Archiv der Deutschen Physikalischen Gesellschaft, Berlin

James **Franck** papers, RLUC, correspondence box nos. 1–10

Walther **Gerlach** papers, DMM, collection 80, correspondence & expert opinions

Friedrich-Adolf **Paneth** papers, AMPG III, 45, correspondence with physicists

Otto **Hahn** papers, AMPG, III, rep. 14, correspondence after 1945

Rudolf **Ladenburg** collection, AMPG Va, rep. 7

Max von **Laue** papers, AMPG, III, rep. 50, correspondence incl. suppl. folder 7/7ff. to Theo von Laue; also DMM, inventory 1976-20

Lise **Meitner** papers, Churchill College, Cambridge, England, correspondence

Sascha **Mogun** papers, UAS, SN 26

Michael **Polányi** papers, RLUC, correspondence from 1933, box nos. 2–5, 11, 16, and notes for broadcast to Germany, Aug. 1945; other ms. in box 30, Berlin visit notebook in box 44

Eugene **Rabinowitch** papers, RLUC, correspondence and "Beginnings of a draft for talk in Germany, which was cancelled, Aug. 1948," box nos. 8–9

Erich **Regener** papers, UAS, SN 16 and the university dossiers UAS 57/117

Ferdinand **Schmidt** papers, UAS 57/411

Arnold **Sommerfeld** papers, DMM, correspondence in inventory 1977-28, a selection of which has meanwhile been edited by M. Eckert and is accessible on the Internet at www.lrz-muenchen.de/~Sommerfeld

Sources for the History of Quantum Physics, microfilm DMM

Harald **Wergeland** papers, Universitetsbiblioteket Trondheim, Norway, box 1 & 7

ABOUT THE AUTHOR

Klaus Hentschel studied physics and philosophy at the University of Hamburg. In 1986 he earned his *Magister* in philosophy and in 1987 a *Diplom* in theoretical high-energy physics; in 1988 he was awarded a DAAD stipend to conduct research at the Collected Papers of Albert Einstein at Boston University. In 1989 he obtained his doctorate in history of science and directly afterwards a Rathenau scholarship of the Verbund für Wissenschaftsgeschichte at the Technische Universität Berlin. From 1991, he was *Assistent* at the Institute for History of Science, University of Göttingen, and submitted his *Habilitation* thesis in the history of science in 1995. For the academic year 1996/97, he was fellow at the Dibner Institute for the History of Science and Technology at MIT, Cambridge, Mass., and 1997–2002 *Oberassistent* (roughly assistant professor) at the Institute for History of Science, University of Göttingen, and then Ernst Cassirer Guest Professor at the University of Hamburg. From 2003 to 2005, he was recipient of a grant for experienced researchers by the Deutsche Forschungsgemeinschaft to write a book at the University of Berne, Switzerland. For the winter term of 2005/06, he was substitute professor at the University of Stuttgart. He received a call to the University of Halle as Lichtenberg professor for comparative history of science and recently accepted a full professorship in the history of science and technology at the University of Stuttgart.

Academic prizes: 1990: Kurt Hartwig Siemers prize of the *Hamburger Wissenschaftlichen Stiftung*; 1992: Heinz Maier-Leibnitz prize of the *Bundesminister für Bildung und Wissenschaft*; 1993: Paul Bunge prize of the Hans R. Jenemann Stiftung, conferred by the *German Bunsen Society* and *Gesellschaft Deutscher Chemiker*; 1998: prix international d'histoire des sciences Marc-Auguste-Pictet, awarded by the Genevan *Société de Physique et d'Histoire Naturelle*; 1999: Leopoldina prize for history of science awarded by the *Deutsche Akademie der Naturforscher Leopoldina*, Halle; 2003: elected member of the *Deutsche Akademie der Naturforscher Leopoldina*.

Published books: *Interpretationen und Fehlinterpretationen der speziellen und der allgemeinen Relativitätstheorie durch Zeitgenossen Albert Einsteins*, Basel: Birkhäuser 1990 (Science Networks, vol. 6); *The Einstein Tower. An Intertexture of Dynamic Construction, Relativity Theory, and Astronomy*, Stanford Univ. Press 1997 (Writing Science series); *Zum Zusammenspiel von Instrument, Experiment und Theorie* (a study on the discovery of solar redshift 1880–1960) Hamburg: Verlag Dr. Kovač, 1998; *Mapping the Spectrum:*

Techniques of Visual Representation in Research and Teaching, Oxford Univ. Press 2002; *Gaussens unsichtbare Hand: Der Universitätsmechanicus und Maschineninspector Moritz Meyerstein, – ein Instrumentenbauer im 19. Jahrhundert*, Göttingen: Vandenhoeck & Ruprecht, 2005. Editor of *Physics and National Socialism. An Anthology of Primary Sources* Basel: Birkhäuser, 1996; Max Planck's letter diary, Berlin: ERS-Verlag 1999.

A complete publications list, including numerous scholarly articles in leading international journals, is accessible at: www.uni-stuttgart.de/hi/gnt/hentschel

ABOUT THE TRANSLATOR

Ann M. Hentschel is an American free-lance translator specializing in the history of science. Translations by her include the correspondence volumes of *The Collected Papers of Albert Einstein* (thus far 1914–1921); the biography by Andreas Karachalios *Erich Hückel – From Physics to Quantum Chemistry*; various essays in *Michael Frayn's Copenhagen in Debate*, edited by Matthias Dörries; the reprint edition of *Albertus Seba: Cabinet of Natural Curiosities 1734–1765*; Jost Lemmerich's extensive catalog for the *Röntgen Rays Centennial* (Würzburg 1995); and Siegfried Grundmann's work, *The Einstein Dossiers. Science & Politics – Einstein's Berlin Period*. She recently authored the geographical biography *Albert Einstein – "Those Happy Bernese Years"* (Berne: Stämpfli, 2005), a historically based tour through the Swiss capital at the turn of the twentieth century.

SUMMARY

Few scientific communities have been more thoroughly studied than 20th-century German physicists. Yet their behavior and patterns of thinking immediately after the war remain puzzling. For the first five years after World War II, German physicists abandoned their old battles (theoreticians vs. experimenters, Berliners vs. non-Berliners, Prussians vs. southerners, university vs. industry, and abstract physics vs. down-to-earth "Aryan physics") and a strange solidarity emerged. Former enemies were suddenly willing to exonerate each other blindly with "whitewash" certificates, and even morally upright physicists who had longed for a rapid end to the Nazi dictatorship began to write tirades against the "denazification mischief" or the "export of scientists." Figures we thought we understood so well (such as Max von Laue and Otto Hahn) suddenly started acting in an incomprehensible way. Former personal idiosyncracies melded into a strangely uniform pattern of rejection or resistance to the occupying Allied powers, with attendant repressed feelings of guilt and self-pity. Politics was once again perceived as a dirty business, far away from their own scientific pursuits. Even those who did not completely suppress feelings of guilt did not speak about it in public because they were afraid any such statement would only be used as backing for even harder sanctions against their discipline.

Using tools from the history of mentality, such as close analysis of serial publications and correspondence of physicists, I analyze several of these tendencies. The perspective of German emigré physicists, as reflected in their confidential letters and some reports about the situation observed from visits back home, will embellish our portrait of this collective with contemporary views from the outside.

BIBLIOGRAPHY

Albrecht, Helmuth (ed.) [1993] *Naturwissenschaft und Technik in der Geschichte*, Stuttgart: GNT-Verlag.

Albrecht, Ulrich, Andreas Heinemann-Grüder & Arend Wellmann [1992] *Die Spezialisten. Deutsche Naturwissenschaftler und Techniker in der Sowjetunion nach 1945*, Berlin: Dietz.

Anonyma [1959] *Eine Frau in Berlin. Tagebuchaufzeichnungen vom 20. April bis 22. Juni 1945*, Geneva: Kossodo (reprint in Frankfurt: Eichborn, 2003).

Anonymous [1947] "The fate of German science. Impressions of a BIOS officer," *Discovery* **8**, no. 8, Aug., pp. 239–243.

Anonymous [1949] "Die 'Unpolitischen'," *Geist und Tat* **4**, p. 229.

Arendt, Hannah [1993] *Besuch in Deutschland*, Berlin: Rotbuch-Verlag.

Ariès, Philippe [1978] "L'histoire des mentalités," in *La nouvelle histoire*, ed. by Roger Chartier & Jacques Revel, Paris: CEPL, pp. 402–423.

Ash, Mitchell G. [1995a] "Verordnete Umbrüche – konstruierte Kontinuitäten: Zur Entnazifizierung von Wissenschaftlern und Wissenschaft nach 1945," *Zeitschrift für Geschichtswissenschaft* **43**, pp. 903–923.

— [1995b] "Wissenschaftswandel in Zeiten politischer Umwälzungen: Entwicklungen, Verwicklungen, Abwicklungen," *NTM* new ser. **3**, pp. 1–21.

— [1999] "Scientific changes in Germany 1933, 1945, 1990: Towards a comparison," *Minerva* **37**, pp. 329–354.

— [2001] "Wissenschaft und Politik als Ressourcen füreinander. Programmatische Überlegungen am Beispiel Deutschlands," in *Wissenschaftsgeschichte heute. Festschrift für Peter Lundgren*, Bielefeld: Verlag Regionalgeschichte, pp. 117–134.

Badash, Lawrence [2003] "From security blanket to security risk: Scientists in the decade after Hiroshima," *History and Technology* **19**, pp. 241–256.

Balfour, Michael & John Mair [1959] *Four-Power Control in Germany and Austria 1945–46*, Oxford Univ. Press, 1956. German translation of part I: *Vier-Mächte-Kontrolle in Deutschland 1945–1946*, Düsseldorf: Droste.

Barth, Karl [1947] "Der deutsche Student," *Die Neue Zeitung*, 8 Dec., no. 98, feuilleton, pp. 3–4.

Becker, Heinrich, Hans-Joachim Dahms & Cornelia Wegeler (eds.) [1998] *Die Universität Göttingen unter dem Nationalsozialismus*, Munich: Sauer, 2nd exp. edn.

Becker, Norbert & Franz Quarthal (eds.) [2004] *Die Universität Stuttgart nach 1945. Geschichte – Entwicklungen – Persönlichkeiten*, Stuttgart: Thorbecke.

Benz, Wolfgang [1990] *Herrschaft und Gesellschaft im nationalsozialistischen Staat. Studien zur Struktur- und Mentalitätsgeschichte*, Frankfurt: Fischer.

— (ed.) [1990] *Deutschland seit 1945. Entwicklungen in der Bundesrepublik und in der DDR*, Bonn: Bundeszentrale für politische Bildung.

— [1996] "Aufklärung als Besatzungszweck. Zur Erneuerung des demokratischen Denkens in Deutschland nach 1945," in: J. Hess (ed.) *Heidelberg 1945*, Stuttgart: Steiner, pp. 82–88.

— (ed.) [1999] *Deutschland unter alliierter Besatzung 1945–1945/55. Ein Handbuch*, Berlin: Akademie-Verlag.

Beyerchen, Alan [1980] *Scientists under Hitler: Politics and the Physics Community in the Third Reich*, New Haven: Yale Univ. Press, 1977; German translation: *Wissenschaftler unter Hitler. Physiker im Dritten Reich*, Cologne.

Beyler, Richard H. [1994] *From Positivism to Organicism: Pascual Jordan's Interpretations of Modern Physics in Cultural Context*, Ph.D. thesis, Harvard Univ. (UMI 9500016).

— [1996] "The concept of specialization in debates on the role of physics in postwar Germany," in *The Emergence of Modern Physics*, ed. by Dieter Hoffmann et al., Pavia, pp. 389–401.

— [2003] "The demon of technology, mass society and atomic physics in West Germany, 1945–57," *History and Technology* **19**, pp. 227–239.

Bird, Geoffrey [1978] "The universities," in: *The British in Germany: Educational Reconstruction after 1945*, ed. by Arthur Hearnden, London: Hamish Hamilton, pp. 146–157.

Birke, Adolf M. [1984] "Geschichtsauffassung und Deutschlandbild im Foreign Office Research Department," in *Das britische Deutschlandbild im Wandel des 19. und 20. Jahrhunderts*, ed. by Bernd Jürgen Wendt, Bochum: Brockmeyer, pp. 171–197.

Bloch, Marc [1928] 'Pour une histoire comparée des sociétés européennes,' *Revue de synthèse historique* **46**, pp. 15–50.

Borkin, Joseph [1990] *Die unheilige Allianz der I.G. Farben. Eine Interessengemeinschaft im Dritten Reich*, Frankfurt am Main: Campus, 3rd edn. (1st edn. 1979).

Bourdieu, Pierre [1993] *Sozialer Sinn. Kritik der theoretischen Vernunft*, Frankfurt: Suhrkamp (1st edn. *Le sens pratique*, Paris: Éditions de Minuit, 1980).

Bower, Tom [1987] *The Paperclip Conspiracy. The Battle for the Spoils and Secrets of Nazi Germany*, London: M. Joseph.

Brink, Cornelia [1998] *Ikonen der Vernichtung. Öffentlicher Gebrauch von Fotografien aus NS-Konzentrationslagern nach 1945*, Berlin: Akademie-Verlag.

Broch, Hermann [1986] *Briefe über Deutschland 1945–1949. Die Korrespondenz mit Volkmar von Zühlsdorff*, Frankfurt: Suhrkamp.

Bröckling, Ulrich [1993] "Zwischen Hitler und Adenauer: Vergessen, Verleugnen, Wegarbeiten in einer Zeit ohne Führer," *Kritik und Krise* **6**, pp. 50–56.

Bruch, Rüdiger vom & Brigitte Kaderas (eds.) [2002] *Wissenschaften und Wissenschaftspolitik. Bestandsaufnahmen zu Formationen, Brüchen und Kontinuitäten im Deutschland des 20. Jahrhunderts*, Wiesbaden: Steiner.

Brüche, Ernst [1946] "Deutsche Physik und die deutschen Physiker," *Neue PB* **2**, pp. 232–236.

— [1947] "NS-Physiker," *PB* **3**, p. 168 (commentary on Stark [1947] on p. 288; see also von Laue [1947]).

Brücher, Hildegard & Clemens Münster [1949] "Deutsche Forschung in Gefahr?," *Frankfurter Hefte* **4**, pp. 333–344.

Brüdermann, Stefan [1997] "Entnazifizierung in Niedersachsen," in *Übergang und Neubeginn. Beiträge zur Verfassungs und Verwaltungsgeschichte Niedersachsens in der Nachkriegszeit*, ed. by Dieter Poestges, Göttingen: Vandenhoeck & Ruprecht, pp. 97–118 (Veröffentl. der Niedersächsischen Archivverwaltung no. 52).

Brumlik, Micha [2001] "Deutschland – eine traumatische Kultur," in: Klaus Naumann (ed.) *Nachkrieg in Deutschland. Die Bundesrepublik und der Nationalsozialismus*, Frankfurt/M., pp. 409–420.

Burke, Peter [1986] "Strengths and weaknesses in the history of mentalities," *History of European Ideas* **7**, pp. 439–451.

Carson, Cathryn [1999] "New models for science in politics: Heisenberg in West Germany," *Historical Studies in the Physical Sciences* **30:1**, pp. 115–171.

— [2002] "Nuclear energy development in postwar West Germany: Struggles over cooperation in the Federal Republic's first reactor station," *History and Technology* 18,3, pp. 233–270.

— [2003] "Bildung als Konsumgut: Physik in der westdeutschen Nachkriegskultur," in Dieter Hoffmann (ed.), pp. 73–85.

— & Michael Gubser [2002] "Science advising and science policy in postwar West Germany: The example of the Deutscher Forschungsrat," *Minerva* **40**, no. 2, pp. 147–179.

Cassidy, David [1992] *Uncertainty: The Life and Science of Werner Heisenberg*, New York: Freeman.

— [1994/96] "Controlling German Science," *Historical Studies in the Physical Sciences* **24** (1994) pp. 197–235, **26** (1996) pp. 193–237.

Deichmann, Ute [2001] *Flüchten, Mitmachen, Vergessen: Chemiker und Biochemiker in der NS-Zeit*, Weinheim: VCH.

— [2002] "Chemiker und Biochemiker in der NS-Zeit," *Angewandte Chemie* **114**, pp. 1364–1383.

"Denkschrift des Forschungsrates" [1949] *GUZ* **4**, no. 18, p. 17 (see also p. 4).

Dinzelbacher, Peter [1993] "Zu Theorie und Praxis der Mentalitätsgeschichte," in *Europäische Mentalitätsgeschichte*, ed. by Dinzelbacher, Stuttgart: Kröner, pp. XV–XXXVII.

Dollinger, Hans & Thilo Vogelsang (eds.) [1967] *Deutschland unter den Besatzungs-mächten 1945–1949*, Munich: Desch.

Dörries, Matthias (ed.) [2005] *Michael Frayn's Copenhagen in Debate. Historical Essays and Documents on the 1941 Meeting Between Niels Bohr and Werner Heisenberg*. Berkeley: Office for History of Science and Technology, Univ. of California.

Dreisigacker, Ernst & Helmut Rechenberg [1994] "50 Jahre Physikalische Blätter," *PB* **50**, pp. 21–23.

— [1995] "Karl Scheel, Ernst Brüche und die Publikationsorgane," *PB* **51**, pp. F135–F142.

Duby, Georges [1961] "Histoire des mentalités," in *Histoire et ses méthodes*, ed. by Charles Samaran, Paris, pp. 937–966 (reprinted in 2 vols., Paris: Gallimard, 1981).

Duerbeck, Hilmar W. & Piotr Flin [2005] "Ludwik Silberstein – Einsteins Antagonist," *Acta Historica Astronomiae* **27**, pp. 186–209.

Eckert, Michael [1990] "Primacy doomed to failure: Heisenberg's role as scientific advisor for nuclear policy in the Federal Republic of Germany," *Historical Studies in the Physical Sciences* **21**, pp. 29–58.

Eibl, Christina [1999] *Der Physikochemiker Peter Adolf Thiessen als Wissenschaftsorganisator (1899–1990): eine biographische Studie*, Ph.D. thesis, Univ. of Stuttgart.

Einstein, Albert [1947] "The military mentality," *The American Scholar* **16**, pp. 353–354.

Engelmann, Bernt [1987] *Wir hab'n ja den Kopf noch fest auf dem Hals. Die Deutschen zwischen Stunde Null und Wirtschaftswunder*, Cologne: Kiepenheuer & Witsch.

Enzensberger, Hans Magnus (ed.) [1990] *Europa in Trümmern*, Frankfurt: Eichhorn.

Evans, Richard C. [1947] "Naturforschung in Deutschland," *PB* **3**, no. 1, pp. 12–14.

Farquharson, John [1997] "Governed or exploited? The British acquisition of German technology, 1945–48," *Journal of Contemporary History* **32**, pp. 23–42.

Fassnacht, Wolfgang [2000] *Universitäten am Wendepunkt? Die Hochschulpolitik in der französischen Besatzungszone (1945–1949)*, Freiburg & Munich: Alber.

Febvre, Lucien [1953/65] *Combats pour l'Histoire*, Paris: Colin, 1953; 2nd edn. 1965.

Feige, Hans-Uwe [1992] "Aspekte der Hochschulpolitik der sowjetischen Militäradministration in Deutschland (1945–48)," *Deutschlandarchiv* **25**, pp. 1169–1180.

— [1994] "Zur Entnazifizierung des Lehrkörpers an der Universität Leipzig," *Zeitschrift für Geschichtswissenschaft* **42**, pp. 795–808.

Forman, Paul [1974] "The financial support and political alignment of physicists in Weimar Germany," *Minerva* **12**, pp. 39–66.

Foschepoth, Josef & Rolf Steininger (eds.) [1985] *Die britische Deutschland- und Besatzungspolitik 1945–1949*, Paderborn: Schöningh.

Frank, Charles (ed.) [1993] *Operation Epsilon: The Farm Hall Transcripts*, Bristol: Institute of Physics Publishing.

Frei, Norbert [1999] *Vergangenheitspolitik. Die Anfänge der Bundesrepublik und die NS-Vergangenheit*, Munich: dtv (1st edn. Beck, 1996).

Friess, Peter & Peter M. Steiner (eds.) [1995] *Forschung und Technik in Deutschland nach 1945*, Munich: Deutsches Museum.

Fritzsch, Robert [1972] "Entnazifizierung. Der fast vergessene Versuch einer politischen Säuberung nach 1945," *Aus Politik und Zeitgeschichte* **B24**, pp. 11–30.

Gerlach, Walther [1980] "Konen, Heinrich," *Neue Deutsche Biographie* **12**, pp. 465–466.

Gimbel, John [1990] *Science, Technology, and Reparations. Exploitation and Plunder in Postwar Germany*, Stanford Univ. Press.

Gödde, Joachim [1991] "Entnazifizierung unter britischer Besatzung," *Geschichte im Westen* **6**, pp. 62–73.

Goenner, Hubert & Giuseppe Castagnetti [1996] "Albert Einstein as a pacifist and democrat during the First World War." *Max Planck Institute for History of Science*, Berlin, preprint no. 35.

Goudsmit, Samuel [1946] "Secrecy or science?," *Science Illustrated* **1**, pp. 97–99; German trans.: "Wissenschaft oder Geheimhaltung,' *PB* **2**, pp. 203–207.

— [1947] *Alsos*, New York: Schuman, reprinted 1983 by Tomash Publishers and 1988 as vol. 1 of the series History of Modern Physics, 1800–1950, AIP.

Grundmann, Siegfried [2005] *The Einstein Dossiers. Science and Politics – Einstein's Berlin Period*, trans. of revised 2nd edn. by A. M. Hentschel, Berlin: Springer.

Grüttner, Michael [2004] *Biographisches Lexikon zur nationalsozialistischen Wissenschaftspolitik*, Heidelberg: Synchron (Studien zur Wissenschafts- und Universitätsgeschichte no. 6).

Gugerli, David et al. [2005] *Die Zukunftsmaschine. Konjunkturen der ETH Zürich 1855–2005*, Zurich: Chronos.

Hahn, Otto [1949] "Antwort an eine Delegation," *GUZ* **4**, no. 12, pp. 2–4 (see also no. 18, p. 9, Murray [1949b], von Laue [1949a]).

— [1951] "Erich Regener und die Max-Planck-Gesellschaft," *Zeitschrift für Naturforschung* **6a**, issue no. 11.

— & F. H. Rein [1947] "Einladung nach USA," *PB* **3**, no. 2, pp. 33–35.

Haritonow, Alexander [1992] "Entnazifizierung an der Bergakademie Freiberg 1945–1948," *Bildung und Erziehung* **45**, pp. 433–441.

Harwood, Jonathan [1993] " 'Mandarine' oder 'Aussenseiter'? Selbstverständnis deutscher Naturwissenschaftler (1900–1933)," in *Sozialer Raum und akademische Kulturen*, ed. by Jürgen Schriewer et al., Frankfurt: Peter Lang, pp. 183–212.

— [2000] "The rise of the party-political professor? Changing self-understandings among German academics, 1890–1933," in D. Kaufmann (ed.), pp. 21–45.

Hauser, Oswald (ed.) [1987] *Das geteilte Deutschland in seinen internationalen Verflechtungen*, Göttingen & Zurich: Muster-Schmidt Verlag.

Hayes, Peter [2004] *Die Degussa im Dritten Reich. Von der Zusammenarbeit zur Mittäterschaft*, Munich: Beck.

Heinemann, Manfred (ed.) [1981] *Umerziehung und Wiederaufbau. Die Bildungspolitik der Besatzungsmächte in Deutschland und Österreich*, Stuttgart: Klett-Cotta.

— [1990] *Hochschuloffiziere und Wiederaufbau des Hochschulwesens in Westdeutschland 1945–1952*. Part 1: *Die Britische Zone*, Hildesheim: Lax, 1990, part 2: *Die US-Zone*, 1990, part 3: *Die französische Zone*, 1991, part 4: *Die sowjetische Besatzungszone*, 2000.

Heinrich, Rudolf & Hans-Reinhard Bachmann [1989] *Walther Gerlach: Physiker–Lehrer–Organisator*, Munich: Deutsches Museum.

Heisenberg, Werner [1948] "Wer weiß, was wichtig wird? Die Notwendigkeit wissenschaftlicher Forschung," *Die Welt*, no. 143, Thurs., 9 Dec., p. 3.

— [1949] "German atomic research," *NYT*, 30 Jan, sect. 4, p. 8.

— [1969] *Der Teil und das Ganze*, Munich: Piper (Translation: *Physics and Beyond*, Harper & Row, 1971).

— & Karl Wirtz [1948] "Großversuche zur Vorbereitung der Konstruktion eines Uranbrenners," in *FIAT Reviews of German Science 1939–1945*, ed. by W. Bothe & S. Flügge, vol. 13, part II, pp. 142–165.

Henke, Klaus-Dieter [1980] "Aspekte französischer Besatzungspolitik in Deutschland nach dem 2. Weltkrieg," in: *Miscellanea. Festschrift für Helmut Krausnick zum 75. Geburtstag*, ed. by Wolfgang Benz et al. Stuttgart: DVA, pp. 169–191.

Hentschel, Klaus [2000] "Heisenberg, German Culture, and other such horrifying things," *Annals of Science* **57**, pp. 301–306 (essay review of Rose [1998]).

— [2005] *Die Mentalität deutscher Physiker in der frühen Nachkriegszeit (1945–1949)*. Heidelberg: Synchron Publishers (Studien zur Wissenschafts- und Universitätsgeschichte no. 11).

— (ed.) & Ann M. Hentschel (ed. asst./trans.) [1996] *Physics and National Socialism. An Anthology of Primary Sources*, Basel: Birkhäuser.

— & Gerhard Rammer [2000] "Kein Neuanfang: Physiker an der Universität Göttingen 1945–1955," *Zeitschrift für Geschichtswissenschaft* **8**, pp. 718–741.

— & Gerhard Rammer [2001] "Physicists at the University of Göttingen, 1945–1955," *Physics in Perspective* **3** pp. 189–209.

— & Monika Renneberg [1995] "Eine akademische Karriere. Der Astronom Otto Heckmann im Dritten Reich," *Vierteljahrshefte für Zeitgeschichte* **43**, pp. 581–610.

Herf, Geoffrey [1997] *Divided Memory. The Nazi Past in the Two Germanys*, Cambridge: Harvard Univ. Press.

Hermann, Armin [1979] *Die neue Physik. Der Weg in das Atomzeitalter*, Munich: Heinz Moos.

— [1993] "Science under foreign rule; policy of the Allies in Germany 1945–1949," in *XVIIIth International Congress of History of Science Hamburg–Munich. Final Report*, ed. by Fritz Krafft & Christoph J. Scriba, Stuttgart: Steiner, pp. 75–86.

— [1995] "Die Deutsche Physikalische Gesellschaft 1899–1945," *PB* **51**, pp. F61–F105.

Hinrichs, Ernst [1979/80] "Zum Stand der historischen Mentalitätsforschung in Deutschland," *Ethnologia Europaea* **11**, pp. 226–233.

Hoffmann, Dieter (ed.) [2002] "Carl Ramsauer, die Deutsche Physikalische Gesellschaft und die Selbstmobilisierung der Physikerschaft im 'Dritten Reich'," in Maier (ed.), pp. 273–304.

— (ed.) [2003] *Physik im Nachkriegsdeutschland*, Frankfurt: Harri Deutsch.

— [2003] "P. Jordan im Dritten Reich – Schlaglichter," *Max Planck Institute for History of Science*, Berlin, preprint no. 248.

— & Thomas Stange [1997] "East-German physics and physicists in the light of the 'Physikalische Blätter'," in *The Emergence of Modern Physics*, ed. by D. Hoffmann et al., Pavia: Goliardica Pavese, pp. 521–529.

— & Rüdiger Stutz [2003] "Abraham Esau als Industriephysiker, Universitäts-direktor und Forschungsmanager," in Hoßfeld et al. (eds.), pp. 136–179.

— & Mark Walker (eds.) [2007] *Physiker zwischen Autonomie und Anpassung. Die Deutsche Physikalische Gesellschaft im Dritten Reich*, Weinheim: Wiley-VCH.

Hool, Alessandra, Kärin Nickelsen & Gerd Graßhoff [2003] *Safe No. 109. Theodore von Kármán – Flugzeuge für die Welt und eine Stiftung für Bern*, Bern Studies in the History and Philosophy of Science.

Hoßfeld, Uwe, Jürgen John, Oliver Lemuth & Rüdiger Stutz (eds.) [2003] *"Kämpfe-rische Wissenschaft" Studien zur Universität Jena im Nationalsozialismus*, Cologne: Böhlau.

Hückel, Ernst [1975] *Ein Gelehrtenleben, Ernst u. Satire*, Weinheim: Verlag Chemie.

Jansen, Christian [1992] *Professoren und Politik: Politisches Denken und Handeln der Heidelberger Hochschullehrer 1914–1935*, Göttingen: Vandenhoeck & Ruprecht (Kritische Studien zur Geschichtswissenschaft, vol. 99).

Jaspers, Karl [1951] "Das Gewissen vor der Bedrohung durch die Atombombe" in: *Rechenschaft und Ausblick. Reden und Aufsätze*, Munich: Piper, 1961, pp. 314–320.

— [1957] *Die Atombombe und die Zukunft des Menschen*, Munich: Piper, 1st edn. subtitled: *Ein Radiovortrag*; 2nd edn. 1958 with new subtitle: *Politisches Bewußtsein in unserer Zeit.*

Jens, Walter [1977] *Eine Deutsche Universität. 500 Jahre Tübinger Gelehrtenrepublik*, Munich: Kindler.

Jessen, Ralph & Jakob Vogel (eds.) [2002] *Wissenschaft und Nation in der europäischen Geschichte*, Frankfurt: Campus.

Jordan, Pascual [1956] *Der gescheiterte Aufstand. Betrachtungen zur Gegenwart*, Frankfurt: Klostermann.

Judt, Matthias & Burghard Ciesla (eds.) [1996] *Technology Transfer out of Germany after 1945*, Amsterdam: Harwood.

Kaiser, Walter [1997] "Wissenschaft und Technik nach 1945," in König et al. (eds.), pp. 241–256.

Kant, Horst [2002] "Werner Heisenberg and the German Uranium Project. Otto

Hahn and the Declarations of Mainau and Göttingen," *Max Planck Institute for History of Science* Berlin, preprint no. 203.

Kármán, Theodore von [1968] *Die Wirbelstraße – Mein Leben für die Luftfahrt*, Hamburg: Hoffmann & Campe.

Kaufmann, Doris (ed.) [2000] *Geschichte der Kaiser-Wilhelm-Gesellschaft im National-sozialismus: Bestandsaufnahme und Perspektiven der Forschung*, Göttingen: Wallstein.

Kellermann, Henry [1978] *Cultural Relations as an Instrument of U.S. Foreign Policy: The Educational Exchange Program between the United States and Germany, 1945–1954*, Washington, D.C.

Kielmansegg, Peter Graf [1989] *Lange Schatten. Vom Umgang der Deutschen mit der nationalsozialistischen Vergangenheit*, Berlin: Siedler.

Kirsten, Christa & Hans-Jürgen Treder (eds.) [1979] *Albert Einstein in Berlin 1913–1933*, Berlin: Akademie-Verlag.

Klaar, Frank [1994] "Die 'Krise' als Gegenstand der Mentalitätsforschung," in *Mentalität und Gesellschaft im Mittelalter. Festschrift für Ernst Werner*, ed. by S. Tanz, Frankfurt, pp. 301–319.

Kleinert, Andreas [1983] "Das Spruchkammerverfahren gegen Johannes Stark," *Sudhoffs Archiv* **67**, pp. 13–24.

— [2002] " 'Die Axialität der Lichtemission und Atomstruktur." Johannes Starks Gegenentwurf zur Quantentheorie,' in *Chemie – Kultur – Geschichte: Festschrift für Hans-Werner Schütt*, ed. by A. Schürmann & B. Weiss, Diepholz & Berlin: GNT-Verlag.

Klemperer, Victor [1957/75] *LTI. Lingua Tertii Imperii. Die Sprache des Dritten Reiches*, Leipzig: Reclam 1975 (reprint of 3rd edn. Halle: Niemeyer, 1957; 1st edn. 1947).

Klessmann, Christoph [1991] *Die doppelte Staatsgründung. Deutsche Geschichte 1945–1955*, Bonn: Bundeszentrale für politische Bildung, 5th expanded edn.

König, Helmut, Wolfgang Kuhlmann & Klaus Schwabe (eds.) [1997] *Vertuschte Vergangenheit. Der Fall Schwerte und die NS-Vergangenheit der deutschen Hochschulen*, Munich: Beck.

Krafft, Fritz [1981] *Im Schatten der Sensation. Leben und Wirken von Fritz Straßmann*, Weinheim: Verlag Chemie.

Krause, Eckhart, Ludwig Huber & Holger Fischer (eds.) [1991] *Hochschulalltag im Dritten Reich: Die Hamburger Universität 1933–1945*, Hamburg & Berlin: Reimer (3 vols.).

Krönig, Waldemar & Klaus-Dieter Müller (eds.) [1990] *Nachkriegs-Semester. Studium in Kriegs- und Nachkriegszeit*, Stuttgart: Steiner.

Kuhlemann, Frank-Michael [1996] "Mentalitätsgeschichte," in: *Kulturgeschichte heute*, Göttingen: Vandenhoeck & Ruprecht, pp. 182–211.

— [2001] *Bürgerlichkeit und Religion. Zur Sozialgeschichte der evangelischen Pfarrer in Baden*, Göttingen: Vandenhoeck & Ruprecht.

Kuiper, Gerard P. [1946] "German astronomy during the war," *Popular Astronomy* **54**, no. 6, pp. 263–287.

Landrock, Konrad [2003] "Friedrich Georg Houtermans (1903–1966): ein bedeutender Physiker des 20. Jahrhunderts," *Naturwissenschaftliche Rundschau* **56**, pp. 187–178.

Latour, Conrad F. & Thilo Vogelsang [1973] *Okkupation und Wiederaufbau. Die Tätigkeit der Militärregierung in der amerikanischen Besatzungszone Deutschlands 1944–1947*, Munich: DVA.

Laue, Max von [1947] "Bemerkung zu der vorstehenden Veröffentlichung von J. Stark," *PB* **3**, pp. 272–273 (commentary on Stark [1947]).

— [1948] "The wartime activities of German scientists," *Bulletin of the Atomic Scientists* **4**, no. 4, p. 103 (and "A reply to Dr. von Laue" and "Comment by the editor," p. 104; German version in *PB* **3** (1947) pp. 424–425).

— [1949a] "Public relations?" *GUZ* **4**, no. 18, pp. 11–12 (comment on Murray [1949b], Hahn [1949]).

— [1949b] "A report on the state of physics in Germany," *American Journal of Physics* **17**, pp. 137–141.

— [1949c] "Amerika und die deutsche Wissenschaft. Gespräch mit Prof. Max von Laue," *Darmstädter Echo*, **5** no. 173, Wed., 27 July, p. 4.

Lefebvre, Georges [1932] *La Grande Peur de 1789* (a) Paris: Colin; (b) "Die Große Furcht von 1789," partial trans. in *Geburt der bürgerlichen Gesellschaft: 1789*, ed. by Irmgard A. Hartig, Frankfurt: Suhrkamp, 1979, pp. 88–135.

LeGoff, Jacques [1974] "Les mentalités. Une histoire ambiguë," in *Faire de l'Histoire*, ed. by J. Le Goff & P. Nora, Paris: Gallimard, vol. 3, pp. 76–94.

— [1983] "Histoire des Sciences et Histoire des mentalités," *Revue de Synthése* **104**, 3rd ser., nos. 111–112.

— [1992] *Geschichte und Gedächtnis*, Frankfurt: Campus (Italian orig., 1977).

Lemmerich, Jost [1982] *Max Born, James Franck: Der Luxus des Gewissens. Physiker in ihrer Zeit*, Berlin: Staatsbibliothek Preußischer Kulturbesitz.

— (ed.) [1998] *Lise Meitner–Max von Laue. Briefwechsel 1938–48*, Berlin: ERS-Verlag.

— [2002]*Aufrecht im Sturm: James Franck 1882–1964*, manuscript July 2002; forthcoming at Stuttgart: GNT-Verlag.

— [2003] *Lise Meitner zum 125. Geburtstag*, Berlin: Staatsbibliothek.

Lifton, Robert Jay [1986] *The Nazi Doctors: Medical Killing and the Psychology of Genocide*, New York: Basic Books.

Litten, Freddy [1992] *Astronomie in Bayern, 1914–1945*, Stuttgart: Steiner.

— [2000] *Mechanik und Antisemitismus. Wilhelm Müller (1880–1968)*, Munich: Institut für Geschichte der Naturwissenschaften (Algorismus, no. 34).

Lübbe, Hermann [1983] "Der Nationalsozialismus im deutschen Nachkriegsbewußtsein," *Historische Zeitschrift* **236**, pp. 579–599.

Magnus, Kurt [1993] *Raketensklaven. Deutsche Forscher hinter rotem Stacheldraht*, Stuttgart: DVA.

Maier, Helmut [2002] " 'Wehrhaftmachung' und 'Kriegswichtigkeit.' Zur Rüstungstechnologischen Relevanz des KWIs für Metallforschung in Stuttgart vor und nach 1945," *Ergebnisse des Forschungsprogramms Geschichte der KWI im NS*, no. 5.

— (ed.) [2002] *Rüstungsforschung im NS. Organisation, Mobilisierung und Entgrenzung der Technikwissenschaften*, Göttingen: Wallstein.

Mandrou, Robert [1989] "L'Histoire des mentalités," *Encyclopedia Universalis*, vol. 11, pp. 479–481.

Marshall, Barbara [1980] "German attitudes to British military government 1945–47," *Journal of Contemporary History* **15**, pp. 655–684.

Martius, Ursula Maria [1947] "Videant consules...," *Deutsche Rundschau* **70**, no. 11, Nov., pp. 99–102 and the commentary: "Die neue Drachensaat," in *Berlin am Montag*, 6. Dec.

Mehrtens, Herbert [1994] "Kollaborationsverhältnisse: Natur- und Technikwissenschaften im NS-Staat und ihre Historie," in *Medizin, Naturwissenschaft, Technik und Nationalsozialismus, Kontinuitäten und Diskontinuitäten*, ed. by Christoph Meinel & Peter Voswinckel, Stuttgart: GNT-Verlag, pp. 13–33.

Merritt, Anna J. & Richard L. Merritt (eds.) [1970] *Public Opinion in Occupied Germany. The OMGUS Surveys 1945–49*, Urbana: Univ. of Illinois Press.

— (eds.) [1980] *Public Opinion in Semisovereign Germany. The HICOG Surveys, 1949–1955*, Urbana: Univ. of Illinois Press.

Meyenn, Karl von (ed.) [1993] *Wolfgang Pauli: Wissenschaftlicher Briefwechsel mit Bohr, Einstein, Heisenberg et al.*, vol. 3: 1940–1949, Berlin: Springer.

Mitscherlich, Alexander & Margarete Mitscherlich [1967] *Die Unfähigkeit zu trauern. Grundlagen kollektiven Verhaltens*, Munich: Piper.

Mogun, Sascha [1955] "In Memoriam.   Erich Regener gestorben," *Naturwissenschaftliche Rundschau* **8**, pp. 205–206.

Muller, Jerry Z. [1996] "How vital was the Geist in Heidelberg in 1945? Some sceptical reflections," in J. Hess (ed.) *Heidelberg 1945*, Stuttgart: Steiner, pp. 197–200.

Murray, R. C. [1949a] "Wissenschaft und Wissenschaftler im heutigen Deutschland," *Forum – Zeitschrift für das geistige Leben in den deutschen Hochschulen* **3**, pp. 169–171.

— [1949b] " 'Social action': Die Diskussion um Wissenschaft und Wissenschaftler im heutigen Deutschland,' *GUZ* **4**, no. 17, p. 13 (reply to Hahn [1949]; see also ibid., no. 18, p. 9 and Laue [1949a]).

Müser, Helmut [1989] "Marianus Czerny," in *Physiker und Astronomen in Frankfurt*, ed. by Klaus Bethge & Horst Klein, Frankfurt: Fachbereich Physik der Johann Wolfgang Goethe-Universität Frankfurt am Main, pp. 144–169.

Nathan, Otto & Heinz Norden (eds.) [1960] *Einstein on Peace* New York: Simon & Schuster.

Niethammer, Lutz [1972] *Entnazifizierung in Bayern. Säuberung und Rehabilitierung unter amerikanischer Besatzung*, Frankfurt: Fischer.

Oexle, Gerhard [1994] "Wie in Göttingen die Max-Planck-Gesellschaft entstand," *Max-Planck-Gesellschaft Jahrbuch 1994*, pp. 43–60.

Oppenheimer, J. Robert [1948] "Die Physik in der gegenwärtigen Welt," *Neue Zürcher Zeitung*, Sat., 22 May, no. 139, sheet 7.

Overesch, Manfred [1992] *Das besetzte Deutschland 1945–1949. Eine Tageschronik der Politik, Wirtschaft, Kultur*, Augsburg: Weltbild-Verlag.

Paneth, F. A. [1948] "Scientific research in the British zone of Germany," *Nature* **161**, pp. 191–192.

Parker, R. A. C. [1987] "British attitudes to Germany 1944–1946," in Oswald Hauser (ed.), pp. 33–44.

Peltzer, Lilli [1995] *Die Demontage deutscher naturwissenschaftlicher Intelligenz nach dem 2. Weltkrieg – Die Physikalisch-Technische Reichsanstalt 1945/1948*, Berlin: ERS-Verlag.

Perron, Oskar [1946] "Verfälschung der Wissenschaften," *Die Neue Zeitung*, **2**, no. 80, 7 Oct., p. 1.

Phillips, David (ed.) [1983] *German Universities after the Surrender – British Occupation Policy and the Control of Higher Education*, Oxford: University of Oxford Department of Educational Studies.

Pilling, Iris [1996] " 'Der fehlende Zorn des Volkes' – Überlegungen Hannah Arendts zur Nachkriegszeit," in J. Hess (ed.), *Heidelberg 1945*, Stuttgart: Steiner, pp. 159–170.

Pingel, Falk [1982] " 'Die Russen am Rhein?' Zur Wende der britischen Besatzungspolitik im Frühjahr 1946," *Vierteljahrshefte für Zeitgeschichte* **30**, pp. 98–116.

Radkau, Joachim [1995] "Kontinuität und Wandel nach 1945 in West- und Ostdeutschland," in: Friess & Steiner (eds.), pp. 57–75.

Rammer, Gerhard [2003] 'Göttinger Physiker nach 1945. Über die Wirkung kollegialer Netze,' *Göttinger Jahrbuch* **51**, pp. 83–104.

— [2004] *Die Nazifizierung und Entnazifizierung der Physik an der Universität Göttingen*, Ph.D. thesis in history of science, Univ. of Göttingen.

Ramsauer, Carl [1947] 'Zur Geschichte der Deutschen Physikalischen Gesellschaft in der Hitlerzeit,' *PB* **3**, pp. 111–114.

— [1949] *Physik–Technik–Pädagogik: Erfahrungen und Erinnerungen*, Karslruhe: Braun.

Rauh-Kühne, Cornelia [1995] "Die Entnazifizierung und die deutsche Gesellschaft," *Archiv für Sozialgeschichte* **35**, pp. 35–70.

Raulff, Ulrich (ed.) [1989] *Mentalitäten-Geschichte. Zur historischen Rekonstruktion geistiger Prozesse*, Berlin: Wagenbach.

Regener, Erich [1947] "Mitverantwortlichkeit der wissenschaftlich Tätigen," *PB* **3**, pp. 169–170.

— [1949] "Denkschrift betreffs Notwendigkeit der Förderung der physikalischen Wissenschaft," supplement to *PB* **5**, issue 4.

Reichardt, Rolf [1979/80] "Für eine Konzeptualisierung der Mentalitätshistorie," *Ethnologia Europaea* **11**, pp. 234–241.

Reid, Constance [1976] *Courant in Göttingen and New York. The Story of an Improbable Mathematician*, New York & Berlin: Springer.

Rein, F. Hermann [1979/80] " 'Entnazifizierung' und Wissenschaft," *GUZ* **1**, no. 1, 11 Dec., pp. 7–9.

Remy, Steven P. [2002] *The Heidelberg Myth. The Nazification and Denazification of a German University*, Cambridge, Mass: Harvard Univ. Press.

Renneberg, Monika & Mark Walker (eds.) [1994] *Science, Technology and National Socialism*, Cambridge Univ. Press.

Ringel, G. K. & Richard Grammel (eds.) [1947] *Denkschrift: Die Technische Hochschule Stuttgart einst und jetzt*, Stuttgart, 21 July (typed with original photographs, in UAS, SA 2/167).

Ringer, Fritz [1969] *The Decline of the German Mandarins: The German Academic Community*, Cambridge: Harvard Univ. Press.

Rompe, Robert [1947] "Die Verfälschungen der exakten Wissenschaften," *Forum* **1**, no. 4, pp. 135–136.

Rose, Paul Lawrence [1998] *Heisenberg and the Nazi Atomic Bomb Project. A Study in German Culture*, Berkeley: Univ. of California Press (see also Hentschel [2000]).

Rubinstein, M. [1947] "Importation of German scientists into America and Britain," *New Times*, no. 10, pp. 16–19.

Rusinek, Bernd-A. [1998] "Schwerte/Schneider: Die Karriere eines Spagatakteurs 1936–1995," in *Der Fall Schwerte im Kontext*, ed. by Helmut König, Opladen: Westdeutscher Verlag.

Sachse, Carola [2002] " 'Persilscheinkultur.' Zum Umgang mit der NS-Vergangenheit in der Kaiser-Wilhelm/Max-Planck-Gesellschaft," in *Akademische Vergangenheitspolitik*, ed. by Bernd Weisbrod, Göttingen: Wallstein, pp. 217–245.

Salomon, Ernst von [1951] *Der Fragebogen*, Reinbek near Hamburg: Rowohlt, 1961 (1st edn. 1951).

Santner, Eric L. [1992] "History beyond the pleasure principle – some thoughts on the representation of trauma," in: Saul Friedländer (ed.) *Probing the Limits of Representation: Nazism and the 'Final Solution'*, Cambridge: Cambridge Univ. Press, pp. 143–154.

Schaper, Michael et al. (eds.) [2002] *Deutschland nach dem Krieg 1945–1955*, Hamburg (Geo Epoche no. 9).

Scherpe, Klaus R. (ed.) [1982] *In Deutschland unterwegs*, Stuttgart: Reclam.

Schirrmacher, Arne [2005a] "Wiederaufbau ohne Wiederkehr: Die Physik in Deutschland in den Jahren nach 1945 und die historiographische Problematik des Remigrationskonzepts" (Arbeitspapier des MZWTG), Munich.

— [2005b] "Drei Männer Arbeit in der frühen Bundesrepublik," *Max Planck Institut für Wissenschaftsgeschichte, Preprints* **296**, Oct.

Schlüpmann, Klaus [2002] *Vergangenheit im Blickfeld eines Physikers (eine Wissenschaftsstudie)*, draft biography of Hans Kopfermann, online Version 18 Jan.

Schmucker, Georg: *Abraham Esau: Eine wissenschaftspolitische Biographie*, Master thesis (Magisterarbeit) Univ. of Stuttgart, 1992.

Schücking, Engelbert L. [1999] "Jordan, Pauli, politics, and a variable gravitational constant," *Physics Today* **52**, no. 10, pp. 26–31.

Schulze, Hagen [1985] "Mentalitätsgeschichte," *Geschichte in Wissenschaft und Unterricht* **36**, pp. 247–270.

Schüring, Michael [2006] *Minervas verstoßene Kinder. Vertriebene Wissenschaftler und die Vergangenheitspolitik der Max-Planck-Gesellschaft*, Göttingen: Wallstein.

Seemann, Silke [2002] *Die politischen Säuberungen des Lehrkörpers der Freiburger Universität nach dem Ende des zweiten Weltkriegs (1945–1957)*, Freiburg: Rombach-Verlag.

Sellin, Volker[1985] "Mentalität und Mentalitätsgeschichte," *Historische Zeitschrift* **241**, pp. 555–598.

— [1996] "Die Universität Heidelberg im Jahre 1945," in J. Hess (ed.) *Heidelberg 1945*, Stuttgart: Steiner, pp. 91–106.

Sime, Ruth Lewin [1996] *Lise Meitner. A Life in Physics*, Berkeley: Univ. of California Press.

— [2006] "The politics of memory: Otto Hahn and the Third Reich," *Physics in Perspective* **8**, pp. 3–51.

Simonsohn, Gerhard [1992] "Physiker in Deutschland 1933–1945," *PB* **48**, pp. 23–28.

Söllner, Alfons (ed.) [1986] *Zur Archäologie der Demokratie in Deutschland. Vol. 2: Analysen von politischen Emigranten im amerikanischen Außenministerium 1946–1949*, Frankfurt: Fischer, esp. pp. 177–249.

Sontheimer, Kurt [1991] *Die Adenauer-Ära: Grundlegung der Bundesrepublik*, Munich: DTV.

Spruch, Grace Marmor [1965] "One man's adventuring in a shiny suit," *The Saturday Review* **48**, II, Dec. 4, pp. 88–90.

Stamm, Thomas [1981] *Zwischen Staat und Selbstverwaltung. Die deutsche Forschung im Wiederaufbau 1945–1965*, Cologne: Verlag Wissenschaft und Politik.

— [1990] "Deutsche Forschung und internationale Integration 1945–1955," in Vierhaus & Vom Brocke (eds.) pp. 886–909.

Stark, Johannes [1947] "Zu den Kämpfen in der Physik während der Hitler-Zeit," *PB* **3**, pp. 271–272 (see also the commentary by Brüche [1947] and the rebuttal by von Laue [1947]).

Steenbeck, Max [1977] *Impulse und Wirkungen*, Berlin: Verlag der Nation.

Stone, Shepard [1947a] "Report on the mood of Germany," *NYT*, 26 Jan., p. 5.

— [1947b] "What to do with Germany: Two newcomers on the bookrack," *NYT*, 7 Dec., pp. 7, 54.

Szabó, Anikó [2000] *Vertreibung, Rückkehr, Wiedergutmachung. Göttinger Hochschullehrer im Schatten des Nationalsozialismus*, Göttingen: Vandenhoeck & Ruprecht.

Tellenbach, Gerd [1974] "Mentalität," in *Geschichte, Wirtschaft, Gesellschaft. Festschrift für Clemens Bauer*, ed. by Erich Hassinger et al., Berlin, pp. 11–30.

Tent, James F. [1982] *Mission on the Rhine. Reeducation and Denazification in American-Occupied Germany*, Univ. of Chicago Press.

— (ed.) [1998] *Academic Proconsul. Harvard Sociologist Edward Y. Hartsthorne and the Reopening of German Universities 1945–46*, Trier: Wissenschaftlicher Verlag.

Thiriet, Jean-Michel [1979/80] "Methoden der Mentalitätsforschung in der französischen Sozialgeschichte," *Ethnologia Europaea* **11**, pp. 208–225.

Treue, Wilhelm [1967] *Die Demontagepolitik der Westmächte nach dem zweiten Weltkrieg*, Göttingen: Niedersächsische Landeszentrale für politische Bildung.

Vierhaus, Rudolf [1983] 'Handlungsspielräume: Zur Rekonstruktion historischer Prozesse,' *Historische Zeitschrift* **237**, pp. 293–309.

— & Bernhard Vom Brocke (eds.) [1990] *Forschung im Spannungsfeld von Politik und Gesellschaft: Geschichte und Struktur der Kaiser-Wilhelm/Max-Planck-Gesellschaft*, Stuttgart: Deutsche Verlagsanstalt.

Vogt, Annette [2002] "Vertreibung und Verdrängung. Erfahrungen von Wissenschaftlerinnen mit Exil und 'Wiedergutmachung" in der Kaiser-Wilhelm/Max-Planck-Gesellschaft (1933–1955),' *Dahlemer Archivgespräche* **8**, pp. 93–136.

Vollnhals, Clemens (ed.) [1991] *Entnazifizierung. Politische Säuberung und Rehabilitierung in den vier Besatzungszonen 1945–1949*, Munich: dtv.

Vovelle, Michel [1979] "Histoire de mentalités – histoire des résistances ou les prisons de la longue durée," *Technologies, Idéologies et pratiques* **1**, pp. 16–43.

— [1985] *Die französische Revolution. Soziale Bewegung und Umbruch der Mentalitäten*, Frankfurt: Fischer (Italian orig. 1979, French trans. by Peter Schöttler).

Walcher, Wilhelm [1995] "Physikalische Gesellschaften im Umbruch," *PB* **51**, pp. F107–F133.

Walker, Mark [1990] "Legenden um die deutsche Atombombe," *Vierteljahrshefte für Zeitgeschichte* **38**, pp. 45–74.

— [1992] "Physics and propaganda: Werner Heisenberg's foreign lectures under National Socialism," *Historical Studies in the Physical Sciences* **22**, pp. 339–389.

— [1993] "Selbstreflexionen deutscher Atomphysiker," *Vierteljahrshefte für Zeitgeschichte* **41**, pp. 520–542.

— [1994] "The nazification and denazification of physics," in *Hochschule und Nationalsozialismus*, ed. by Walter Kertz, Braunschweig: Universitätsbibliothek der TH Braunschweig, pp. 79–89.

— (ed.) [2003] *Science and Ideology. A Comparative History*, London: Routledge.

— [2006] "Otto Hahn: Responsability and repression," *Physics in Perspective* **8**, pp. 116–163.

Wallach, Curt [1946] "Völkische Wissenschaft – Deutsche Physik," *Deutsche Rundschau* **63**, pp. 126–141.

Weart, Spencer R. [1988] *Nuclear Fear. A History of Images.* Cambridge, Mass.: Harvard Univ. Press.

Wein, Martin [1996] 'Carl Friedrich und Richard von Weizsäcker,' in *Deutsche Brüder: Zwölf Doppelportraits*, Reinbek near Hamburg: Rowohlt, pp. 366–393.

Weisbrod, Bernd [2002] *Akademische Vergangenheitspolitik*, Göttingen: Wallstein.

von Weizsäcker, Carl Friedrich [1948] *Die Geschichte der Natur. Zwölf Vorlesungen.* Göttingen: Vandenhoeck & Ruprecht (also Leipzig: Hirzel).

— [1991] "Ich gebe zu, ich war verrückt," *Der Spiegel* **45**, no. 17, pp. 227–238.

Welsh, Helga [1985] "Die Entnazifizierung der Universität Leipzig. Ein Bericht des Rektors Bernhard Schweizer vom Anfang 1946," *Vierteljahrshefte für Zeitgeschichte* **33**, pp. 339–372.

Wischnath, Johannes Michael [1998] "Eine Frage des Stolzes und der Ehre. Die politische Säuberung der Universität Tübingen und ihr letzter NS-Rektor Otto Stickl," in: *Persilschein, Käferkauf und Abschlachtprämie. Von Besatzern, Wirtschaftswunder und Reformen im Landkreis Tübingen*, pp. 103–123.

Wolfschmidt, Gudrun [1992] "Kiepenheuers Gründung von Sonnenobservatorien im Dritten Reich," *Deutsches Museum, Wissenschaftliches Jahrbuch 1992*, pp. 283–318.

Wolgast, Eike [2001] *Die Wahrnehmung des Dritten Reiches in der unmittelbaren Nachkriegszeit (1945/46)*, Heidelberg: Winter (Schriften der Philosophisch-historischen Klasse der Heidelberger Akademie der Wissenschaften, no. 22).

Zeitz, Katharina [2006] *Max von Laue (1879–1960). Seine Bedeutung für den Wiederaufbau der deutschen Wissenschaft nach dem zweiten Weltkrieg*, Stuttgart: Steiner.

Zuckmayer, Carl [2002] *Geheimreport*, Göttingen: Wallstein.

Zur Nieden, Susanne: *Alltag im Ausnahmezustand. Frauentagebücher im zerstörten Deutschland 1943–45*, Berlin: Orlanda, 2002.

NAME INDEX

This index lists the names of all persons mentioned or quoted in the narrative. Composite terms incorporating proper names (e.g., Max Planck Society) are excluded. Emphasis indicates that additional biographical information (living dates) is provided on the relevant page. An f. following a page number signifies that and the following page; ff. includes the subsequent page. For ranges exceeding three successive pages the first and last page are indicated. Initials occur only where a duplicate surname is separately cited.